Essay on the Geography of Plants

Essay on the Geography of Plants

ALEXANDER VON HUMBOLDT
AND AIMÉ BONPLAND

Edited with an Introduction by Stephen T. Jackson
Translated by Sylvie Romanowski

THE UNIVERSITY OF CHICAGO PRESS :: CHICAGO AND LONDON

The University of Chicago Press, Chicago 60637
The University of Chicago Press, Ltd., London
© 2009 by The University of Chicago
All rights reserved. Published 2009.
Paperback edition 2013.
Printed in the United States of America

22 21 20 19 18 17 16 15 14 13   1 2 3 4 5

ISBN-13: 978-0-226-36066-9 (cloth)
ISBN-13: 978-0-226-05473-5 (paper)
ISBN-13: 978-0-226-36068-3 (e-book)
10.7208/chicago/9780226360683.001.0001

Library of Congress Cataloging-in-Publication Data

Humboldt, Alexander von, 1769–1859.
  [Essai sur la géographie des plantes. English]
  Essay on the geography of plants / Alexander von Humboldt and
Aimé Bonpland ; edited with an introduction by Stephen T. Jackson ;
translated by Sylvie Romanowski.
      p. cm.
  Includes bibliographical references.
  ISBN-13: 978-0-226-36066-9 (cloth : alk. paper)
  ISBN-10: 0-226-36066-0 (cloth : alk. paper)
  1. Phytogeography. 2. Plant ecology. 3. Physical geography.
I. Bonpland, Aimé, 1773–1858. II. Jackson, Stephen T., 1955–
III. Romanowski, Sylvie. IV. Title.

  QK101.H9313 2009
  581.9—dc22

                                    2008038315

♾ This paper meets the requirements of ANSI/NISO Z39.48-1992
(Permanence of Paper).

Contents

Preface

This project had its origins in a chance conversation during a Chicago-to-Paris flight in the summer of 2003. Sylvie Romanowski and I were seated next to each other, and, upon chatting, we learned that we were respectively professors of French literature and botany. I mentioned my frustration in attempting to read a recently acquired reprint volume of Humboldt's *Essai sur la géographie des plantes*. I was not sure whether my problems had to do with archaic idioms in the document or the sad state of my French-language skills after decades of neglect. Sylvie suggested she might be able to help and asked me to send a copy to look at. Upon my return to Laramie, I sent a copy of the *Essai* text to Sylvie. She was intrigued, and upon seeing a facsimile of Humboldt's color plate that accompanied the *Essai*, she developed a definitive case of the "Humboldt virus."

The Humboldt virus is an easy one to catch. All it takes is a reading of part of Humboldt's *Personal Narrative*, or an essay from his *Views of Nature*, or simply a perusal of the Chimborazo profile which accompanied the *Essai*. You are confronted with a man who was interested in nearly everything, who could speak and write authoritatively about the electrical properties of muscles, the philology of the ancient Incas, the political economy of Mexico, and the mineralogy of the Urals; who was equally comfortable conversing with Gauss, Goethe, and Gay-Lussac (not to mention Thomas Jefferson, Tsar Nicholas I, Abraham Mendelssohn-Bartholdy, an anonymous, pedantic poison-master in the upper Orinoco, or Carlos del Pino, the Guayqueria whom Humboldt enlisted as his assistant in Venezuela); who dined on ants and monkeys in the Orinoco and foie gras and caviar in Paris salons; who was passionate in his love of political freedom and his hatred of slavery; who envisioned how hundreds of point-observations of temperature, magnetism, or plant form could be assimilated to reveal global patterns; who led the funeral procession for the fallen Berlin revolutionaries of 1848 at the same time he was serving as chamberlain and confidant to the Prussian king; who was happy to do chemical experiments in a flooded Paris basement, inventory plants while climbing the slopes of Cotopaxi, make exacting astronomical measurements in a mosquito-infested jungle, or dissect electric eels with a local savant in a dusty llanos village; who, when told he must ride across the Quindiu Pass aboard a sillero—a mulatto man saddled with a chair—strapped one of the chairs to his own back and insisted that he carry the sillero; who inspired the scientific career of Charles Darwin, the artistic career of Frederic Edwin Church, and the political career of Simón Bolívar; who was eulogized in America by Ingersoll, Emerson, and Agassiz (respectively an agnostic, a mystic, and a zealot); who was the subject of the dedication page of Edgar Allan Poe's last major work; who spent his considerable inheritance self-funding his scientific explorations, underwriting his scientific monographs, and supporting his less fortunate colleagues; who made substantive contributions to nearly every branch of the natural and social sciences of his time; whose work laid the foundations for a dozen disciplines, ranging from geophysics to biogeography to political economy.

Resistance is pointless once you have begun to engage Humboldt. Though his personality is often remote (his Personal Narrative is anything but personal), his insatiable curiosity and intellectual power are always close at hand. Within a few months of our first meeting, Sylvie and I had agreed to work together on a complete translation of Humboldt's Essai, and before the project was long underway we were both inspired to write accompanying essays.

The Essai sur la géographie des plantes, with its accompanying Tableau physique des Andes et Pays voisins, is one of Humboldt's most influential works. It has long been regarded as one of the foundation texts of ecology and biogeography, but it is more than that. It is the first mature, integrated statement of Humboldt's view of a unified nature, with diverse properties showing coherent patterns in space at local to global scales. The Essai (and its German equivalent, Ideen zu einer Geographie der Pflanzen) has been reprinted at various times since its first publication in 1807. But it has never appeared in an English translation.

Since the Second World War, English has emerged as the lingua franca of science. As a consequence, native-born English speakers working in the sciences no longer have to develop fluency in French, German, or other languages. In fact, scientists at universities in the United States often view foreign languages as a burden, discourage students from taking them, and even try to purge them from the required curriculum. It is an unfortunate fact that most scientists in the United States and other English-speaking nations are unable to read and appreciate Humboldt's Essai.

The primary purpose of this volume, then, is to bring Humboldt's Essai to an English-speaking audience at relatively low cost. But why should anyone in the early twenty-first century read a 200-year-old scientific work, however canonical? The question is obviously easy to answer for historians or humanists, but why should scientists interrupt their busy lives to read Humboldt, or any of the old masters?

Scientists have a well-deserved reputation for being focused on the "here and now"—the burning questions that are at the frontier of knowledge. They tend not to be concerned with the questions and concepts of a decade ago, let alone a century. This is paradoxical, because science is inherently a historical enterprise. The body of current knowledge is built on years, decades, or centuries of previous scholarship and research, and it is subject to the same kinds of historical contingencies and artifacts as any other human endeavor. Some scientists recognize this, and acknowledge the need to step back and examine where a question or concept has come from, how it has evolved, and whether something important has been overlooked along the way. At its most fundamental level, this is good scholarship and leads to a healthy sense of humility. Furthermore, concepts tend to evolve in time, and so confusion and pointless controversy can be avoided by looking at their lineages. And occasionally, long-neglected ideas are rediscovered and brought back to life, helping to solve current puzzles or opening new avenues for discovery.

Historical examination is particularly critical for ecology, evolution, and biogeography, because these are inherently historical sciences. Obviously so, in the sense that the phenomena that they attempt to explain—the abundance and distribution of organisms across the globe, the diversity of life at local to global scales, the origin of features of organisms that fit them to their respective environment, the movement of energy and materials in space and time—all have historical components. None can be fully explained without some knowledge of history, because different processes occur at different rates, Earth's environment changes through time, and system states at any given time are contingent on previous states and events.

However, ecology, evolution, and biogeography are historical in another sense: virtually all of the core concepts date back to the nineteenth century or before, and ideas proposed one or two or even three centuries ago remain relevant and topical today. This undoubtedly arises from the sheer diversity and complexity of the subject material. Ecologists and evolutionary biologists are probably no more susceptible than physicists and chemists to Francis Bacon's "Idols of the Theatre" (i.e., unquestioned paradigms and notions inherited from previous generations), nor are they necessarily more prone to equivocation in resolving important issues. Issues can and do get resolved, and paradigms are subjected to scrutiny. But the nature of the subject matter makes these fields susceptible to erroneously rejecting real mechanisms and processes, and to overestimating the importance or universality of others. Definitive falsification of a mechanism in one case, for one class of organisms at one spatial scale and one temporal scale, provides no insurance that the mechanism is not important in some other case or at some other scale. The historical, contingent nature of the phenomena under investigation compounds this problem.

Thus, ecologists, biogeographers, and evolutionary biologists can still profit from reading classical texts. What Charles Darwin said about animal behavior, ecological competition, or natural variation has more to say to us today than what, for example, Joseph Priestley said about phlogiston or James Clerk Maxwell said about the ether. In ecology and evolution, we live with the legacy of Darwin's emphasis on competition, and many are re-reading Darwin, Wallace, Lyell, and others to see how we got where we are today, and to examine whether we got on a wrong track a long time ago.

There is yet another reason to read Humboldt today. He worked in a time of widespread revolution, repression, and seemingly endless war. Science and scientists have never been immune to larger political, economic, and societal currents. In the United States and most of the rest of the western world in the second half of the twentieth century, science benefited from a liberal political culture that was largely willing to let science run its course and even to embrace the application of scientific knowledge to policy. A prime example is the blossoming of the environmental sciences and their direct incorporation into environmental policy and resource management.

There is no guarantee that the good times for science will continue, however. I write at a time when forces of political reaction have been ascendant in the United States and elsewhere. No part of our society or culture, including science, has been unaffected. From stem-cell research to teaching of evolution to application of climate-change science and conservation biology, powerful political and economic forces are bearing down to subvert, divert, or simply shout down science, its practitioners, and its products. This is bad for science and society. Good science can flourish only in an open society, and as governments become more secretive and authoritarian, and societies more close-minded and self-absorbed, science will suffer. Conversely, science, with its dynamism, skepticism, and progressivism, is essential for a healthy democracy. The one requires the other, and erosion of one will degrade the other. It should be no surprise that many of the leaders of the Enlighten-ment and post-Enlightenment movements for liberal political democracy were also scientists themselves.

Alexander von Humboldt and his contemporaries can provide solace and inspiration in what may appear to be a darkening time. I believe they rep-resent the best our culture has to offer: on the one hand, a commitment to applying our senses and minds systematically toward understanding nature and to applying scientific knowledge to help advance society and enrich people's lives, and, on the other, a commitment to an open, democratic, and egalitarian society. They saw triumphs as well as tragedies in their lifetimes, just as we have seen great successes and grievous failures in human prog-ress, social justice, and environmental health in ours. Many of Humboldt's scientific contemporaries suffered, and some died, for their thoughts and deeds. In our time, at least in the United States and Western Europe, the costs of free thought and speech have not been so high. We can only hope this will not change.

Humboldt spent the last ten years of his life under police surveillance, with all of his correspondence opened and inspected by government cen-sors. Only his long-standing association with the King of Prussia prevented worse treatment. In an 1849 letter to his close friend, the physicist François Arago, Humboldt reflected on the failed political revolutions of 1848: "1849 is the year of reaction. I have saluted 1789 . . . and now, at the age of eighty, I am reduced to the worn-out hope that the fine and ardent wish for free institutions is maintained in the people, that, periodically, it may appear to be asleep but that it is eternal as the electromagnetic storm which sparkles in the sun."[1]

The intervening century-and-a-half suggest that Humboldt's "worn-out hope" was at least partly justified. The chattel slavery Humboldt so despised was outlawed worldwide by the late nineteenth century, universal suffrage and liberal democracy were established over most of Europe, the Ameri-cas, and other parts of the world in the second half of the twentieth century (though not without violent reversals or resistance in many countries), and a concept of universal human rights has been made explicit in the United

1. From a letter dated Potsdam, 9 November 1849, translated and quoted in L. Kel-lner, *Alexander von Humboldt* (Oxford: Oxford University Press, 1963), 218. An alternative translation is provided in E. R. Brann, *The Political Ideas of Alexander von Humboldt* (Madi-son, Wis.: Littel, 1954), 30.

Nations Charter and given at least nominal support by governments in much of the world.

It is my hope that publication of this important work of Humboldt's will lead not only to a greater appreciation of the scholarly debt nearly all earth, ecological, and environmental scientists owe to Humboldt but also to a renewed commitment among scholars to Humboldt's dual vision of scientific pursuit as a key element of an open society and an open society as an essential support for continued scientific progress.

Stephen T. Jackson

Note to the Reader

This book represents a collaborative effort, involving the editor, an environmental scientist with a longstanding interest in the history of science, and the translator, a scholar of seventeenth- and eighteenth-century French culture with a longstanding interest in the natural sciences. We like to think that Alexander von Humboldt, whose interests went far beyond science to embrace history, art, and aesthetics, would have appreciated this joint effort of scholars from distant disciplines who found common ground in his work.

The volume opens with an introductory essay by the editor that provides background and context for the contemporary reader. The essay presents his perspective as a twenty-first century scientist on Humboldt's *Essay on the Geography of Plants* and its current relevance to science and society. We hope it will be useful to orient readers as they approach a scientific work written for an early nineteenth-century audience.

Humboldt's translated text follows in its entirety, preceded by a note from the translator. The text is organized into three parts: a brief "Preface," an "Essay on the Geography of Plants," and a much longer "Physical Tableau of the Equatorial Regions." The latter is a commentary on a color plate, the *Tableau physique des Andes et Pays voisins*, that Humboldt prepared for the volume. In developing this project, the editor and translator agreed at the outset that we should strive to keep as closely as possible to Humboldt's original text. The editor's role was to resolve scientific and other uncertainties in the translation, ensure scientific accuracy and integrity, and make suggestions on clarity and economy. Some names of plants, animals, and people appear to be misspelled in the original work, and some appear in multiple spellings. We made no attempt to correct these apparent errors. Many spellings of Latin plant names, though in error today, are in accord with usage during Humboldt's time. Similarly, we made no corrections to omissions and other apparent errors in the data tables; other than French-to-English translations, numbers and nouns in the tables are exactly as Humboldt represented them.

The lodestar of Humboldt's *Essay* is the large color plate, *Tableau physique des Andes et Pays voisins*, with its iconic profile of Chimborazo. A full-size reproduction of this figure is provided with the volume (see pocket on rear cover). This figure should be consulted while reading the text; as will become apparent to the reader, this was Humboldt's intention. We have provided a translation of the text portions of the *Tableau physique*, which immediately follow the *Essay* translation.

The translations are followed by an analysis of the *Essay* and the *Tableau physique*, written by the translator, from a literary, cultural, and artistic point of view. This essay focuses on the *Tableau physique* as an aesthetic object and then situates Humboldt's pictorial science in the context of the art, science, and philosophy of his time, particularly in France. Writing in French for an audience of French savants, Humboldt prepared the *Essay* text and the plate in Paris, and the work was heavily influenced by French intellectual culture.

The volume concludes with four supplementary sections, prepared by the editor. A tabulation of all the plant names mentioned in the *Essay*, with common names, modern synonymies, and plant families, is intended to aid the reader in negotiating a work that cites specific plants from all over the world, often with archaic names. A brief essay discusses the instruments that accompanied Humboldt in his New World travels, and which underpin the *Tableau physique* and much of the *Essay* text. A series of biographical sketches includes most of the individuals mentioned by Humboldt in the *Essay* as well as a few others who were important to Humboldt's work. Largely forgotten today, they include some of the most important scientific figures of their times. Their lives inform us about the diverse origins of the natural sciences and about the challenges faced by early naturalists and scientists. Finally, a bibliographical essay discusses English-language sources of information on Humboldt, his contemporaries, and his contributions, as well as other sources useful in preparing the volume.

It has been our pleasure and privilege to engage with the thinking of Alexander von Humboldt while pursuing this project. We hope this volume gives twenty-first-century readers an appreciation for this important figure. His silent legacy persists throughout the earth, environmental, and ecological sciences. It is fitting, some two hundred years after Humboldt published his *Essay*, to recognize his role in creating the intellectual world we inhabit.

Stephen T. Jackson, Editor
Sylvie Romanowski, Translator

A Note on Nomenclature

Throughout the accompanying essays we refer to Humboldt's original work as the *Essay*. We have retained the original French *Tableau physique* to refer to the color plate, which is reproduced here in its original language. An English translation of the text of that document is provided separately.

Humboldt's text is divided into three sections: "Préface," "Essai sur la géographie des plantes," and "Tableau physique des régions équatoriales." Throughout the essays and appendices, these sections are referred to as "Preface," "Essay," and "Physical Tableau" respectively. The section titles are placed in quotation marks, and the work as a whole as well as the title of the plate are in italics.

A Note on Units

Humboldt uses two systems of measurement units throughout the *Essay*: the newly developed metric system and the system of measures used during the *ancien régime*. The metric system, with its meters, grams, and degrees Centigrade, was developed during the revolutionary period in France, and it remains unchanged excepting some minor refinements and standardizations. It has been adopted as the universal system for every nation in the world except the United States and a handful of others. It is universally used in scientific work. The older system was still in use when Humboldt's work was written, but disappeared during the mid-nineteenth century, replaced by the metric system.

The standard for linear measurement under the older system was the *toise*, equivalent (at least in Paris) to 1.949 meters. The toise was divisible into six feet, each in turn divided into twelve *pouces* (analogous to the English inch). Each *pouce* was in turn divided into twelve *lignes* or lines, each equivalent to 2.256 millimeters. We have retained Humboldt's original units throughout the translation.

Humboldt expresses temperature using two scales, the Celsius scale (°C), similar to that in use today, and the now-obsolete Réaumur scale (°R), which defines the freezing point of water at 0 and the boiling point at 80.

STJ
SR

Acknowledgments

It has been a rare pleasure to work on this project with Sylvie Romanowski. She shared my enthusiasm for the project from the very beginning and stimulated me to think about science and its history in unforeseen ways. Christie Henry at the University of Chicago Press also believed in the project from our first conversation and has been a steady source of advice and encouragement.

I am grateful to my colleagues Dennis Knight, Jeff Lockwood, Carlos Martinez del Rio, and Bill Reiners, who provided discussions and critiques, and to John Logan Allen, Biff Bermingham, Mark Bush, Mark Lomolino, Roger Williams, and Miles Silman, who all encouraged and informed me at various stages of the project.

Several colleagues kindly advised me on synonymies and other aspects of plant species discussed in the Humboldt *Essay*: Greg Brown and Ron Hartman (vascular plants), Steve Miller (fungi), Bruce McCune (lichens), and Ray Stotler (bryophytes).

Funding for the printing of the color *Tableau physique* was provided by the International Biogeography Society and by several entities at the University of Wyoming (College of Arts and Sciences, Office of Academic Affairs, Department of Botany, and Program in Ecology).

My contributions to this volume were largely accomplished during stray hours stolen on evenings, weekends, and holidays over the past few years. These hours represent time not spent with my wife, Anne Bowen, and our horses. The horses haven't cared; what is recreation for me is work for them. Anne has indulged my passion for Humboldt and history of science, as well as my ever-expanding library. As an academic and a psychologist, she knows something about the nature of obsession.

Stephen T. Jackson

I wish to express my gratitude to several people and institutions that have facilitated this work: the staff at the Muséum national d'Histoire naturelle (Paris), the Library of the Musée des arts et métiers (Paris), the staff of the

Special Collections of the Northwestern University Library (Evanston), especially Russell Maylone and Sigrid Pohl Perry. At the University of Chicago Press Christie Henry has been an unfailing, understanding, and enthusiastic supporter of this project. I thank the University Research and Grants Committee of the Northwestern University Graduate School for its financial aid. I wish to recognize some useful conversations with Sean O'Driscoll, and to thank Holly Woodson Waddell and especially Claire Carpenter for their readings of earlier versions of my essay. Finally, I express my gratitude to Stephen T. Jackson, the initiator of this project, for his collaboration as the co-author of this volume. He not only provided me with much needed information about scientific details about instruments, plants, and animals, he also engaged in an ongoing dialogue that was essential to reaching our common goal: translating Humboldt in such a way as to make his text accessible to a contemporary audience of both scientists and humanists while remaining faithful to his epoch and spirit.

Sylvie Romanowski

Introduction:
Humboldt, Ecology, and the Cosmos

Stephen T. Jackson

I

"I formerly admired Humboldt, I now almost adore him. . . ."[1]

"Humboldt had a direct influence on a wider range of people than any scientist since Newton."[2]

We live in a world shaped intellectually by two nineteenth-century figures, Alexander von Humboldt and Charles Darwin. Humboldt's death in May 1859 came only six months before publication of Darwin's *On the Origin of Species*. Today, nearly 150 years later, virtually everyone in the western world knows something about Darwin while few know anything about Humboldt. Most laypeople know that Darwin had something to do with evolution, and most biologists and environmental scientists are familiar with Darwin's theory of evolution by means of natural selection, his voyage aboard *H.M.S. Beagle*, his fascination with variation and selection in domesticated pigeons and other animals, his disputes with Agassiz and Owen, his principled conduct in sharing credit with Alfred Russel Wallace, and his friendship with Huxley, Hooker, and Lyell. Dozens of Darwinian legends permeate the culture of modern biology, multiple biographies of Darwin and his contemporaries have been published, and most of his works remain in print. Humboldt, in contrast, is hardly a household name today, although a few geographic features—a river in Nevada, a current in the Pacific, some townships and counties scattered about the United States—bear his name. Most of his

---

1. Charles Darwin, letter to the Rev'd. Prof. John Stevens Henslow, from Rio de Janeiro, 18 May 1832. Reprinted in Barlow 1967, 55.

2. Lewis Pyenson and Susan Sheets-Pyenson 1999, 259.

works are long out of print, no biographies have been published since the 1970s, and most scientists have at best a vague idea of the breadth and depth of his accomplishments.

The situation was inverted in 1859. Charles Darwin was well known among natural historians for his work on the systematics of cirripedes,[3] his theories of coral-reef formation, and his observations and collections on the *Beagle* voyage. Although held in high repute by naturalists in England and abroad, his work was largely unknown to the public except for his travel narrative, which had been a minor bestseller in Britain in the 1840s. Humboldt, on the other hand, was an international figure, whose *Cosmos* had been published in eleven languages and was a standard fixture in lay libraries, along with his *Views of Nature* and his *Personal Narrative of a Voyage to the Equinoctial Regions*. Humboldt was the leading scientist and scholar of his day. It is difficult today to comprehend his popular stature—there has been no comparable figure in my lifetime (perhaps a blend of Carl Sagan, Stephen Jay Gould, and Stephen Hawking with some Neil Armstrong, Noam Chomsky, and Edmund Hilary thrown in for good measure). Laypeople and scientists alike read his works and knew of his contributions, and multiple biographies and hagiographies were published in his lifetime. His adventures in South America and elsewhere were widely known; his climbing record on the Chimborazo volcano was not to be broken for another twenty-one years. Humboldt inspired countless individuals to take up the pursuit of science and natural history; had Darwin not read Humboldt's *Personal Narrative* as a Cambridge undergraduate, he would not have embarked shortly thereafter on H.M.S. *Beagle* and would today be considered at most a minor figure in Victorian natural history.[4]

Even the scientists who are most indebted today to Humboldt know relatively little of his work. Ecologists and biogeographers may know that he worked out some general principles of plant geography and drew an analogy between latitudinal and altitudinal zonation of vegetation. Climatologists may be aware that he fostered development of the first international net-

3. Commonly known as barnacles. Darwin devoted eight years, from 1847 to 1854, to a detailed study of the systematics and morphology of barnacles, culminating in a two-volume monograph. During this period, one of his young sons asked a friend who was visiting, "Where does your father do his barnacles?"

4. The pivotal influence of Humboldt's *Personal Narrative* on Darwin's decision to travel to the tropics and pursue the *Beagle* opportunity is amply documented in Darwin's autobiography and correspondence, and in various Darwin biographies (Bowler 1990, Desmond and Moore 1991, Browne 1995).

works of meteorological observations and that he invented the isotherm as a means of describing temperature variation in space. Geologists may know that he advocated the volcanic origin of basalt, and geophysicists may know that he pioneered the field of geomagnetism. Physiologists may know of his experimental work on electricity and muscle function. Tropical botanists may know of the thousands of plant species he described with Bonpland and Kunth.[5] Social and economic geographers may credit Humboldt with developing their field and doing exemplary work on the political economy of Mexico and Cuba. But few geographers will know of Humboldt's contributions to physiology, few geophysicists will know of his work in economic geography, few physiologists will know of his work in atmospheric physics, and few ecologists will know of his work in systematic botany.

Humboldt's voyage to the Americas with Aimé Bonpland between 1799 and 1804 was the defining event of his scientific career. At the time of their departure, Humboldt was a well-established scientist of thirty, with publications in botany, physiology, and mineralogy to his credit. However, his five years in the Canary Islands, Venezuela, Cuba, Mexico, the United States, and the Andes of Colombia, Peru, and Ecuador led to his most important and influential scientific works, and his popular writings and lectures on his travels (which overlapped with the technical ones) established his public renown.

After his return from the New World, Humboldt spent the next twenty-two years in Paris, exhausting his considerable inherited wealth in preparing and publishing the results of his explorations. These included sixteen volumes on botany (including descriptions of some 8,000 plant species, 4,000 of them new), two volumes of zoology and anatomy (primarily descriptions of new animal species), four volumes of astronomical and geophysical observations, two volumes each of geographical and pictorial atlases, three volumes on the history of exploration and nautical astronomy of the New World, four volumes on the political economy of New Spain (Mexico), seven volumes of the never-finished *Personal Narrative of Travels to the Equinoctial Regions of the New Continent*, a volume of popular essays—and one small volume on the geography of plants.

Although the *Essay on the Geography of Plants* was among the shortest

5. Biographical sketches of Bonpland and several dozen other individuals mentioned in Humboldt's *Essay* are provided near the end of this volume. I also include sketches of Kunth and a few other people not mentioned in the *Essay* who played important roles in Humboldt's thinking and writing.

works Humboldt produced, it may have been the most influential, at least scientifically. The Essay, together with its pictorial representation of physical, ecological, and societal properties arrayed along an elevational gradient, changed the way western culture viewed the world. Science was provided a new lens—a geographic lens in which diverse phenomena could be seen to covary systematically across the face of the earth. In previous centuries, Galileo's telescope and Leeuwenhoek's microscope made people aware of worlds beyond previous imagining. Humboldt's more abstract lens made people aware of a world that already lay before their eyes, a world in which the local details of climate, flora, fauna, soil, and culture could all be seen as parts of broader regional and global patterns, just as individual tiles in a mosaic form patterns when viewed in the context of surrounding tiles.

Humboldt's Personal Narrative stimulated people to dream of the tropics, and it motivated a few of them—notably Darwin and Wallace—to go there. Humboldt's Essay provoked people to think about the globe in fundamentally new ways—as a single entity with interlinked biological, physical, and cultural properties varying latitudinally and altitudinally in a systematic and comprehensible fashion. It was the most synthetic and widely influential of the many technical works he published following his voyage. In the Essay, Humboldt integrated a wide array of disparate measures—not only vegetation composition and form, but temperature, geology, atmospheric pressure, atmospheric chemistry, the blueness of the sky, humidity, agricultural practices—into a single view, showing how they varied systematically with altitude and how they were interlinked. Humboldt was aiming for a unitary vision of the world and its phenomena. This was a lifelong pursuit, beginning as early as 1793 and culminating in Cosmos, his five-volume magnum opus, which he was still working on at the time of his death at the age of 90. The Essay constitutes the first full articulation of Humboldt's broad, unitary scientific vision. It was the first substantial work Humboldt published after his return to Europe.

The central ecological and geographical ideas in Humboldt's Essay are not original, in the same sense that little of scientific progress is truly original—to paraphrase Newton, we see further by standing on the shoulders of giants.[6] Botanical, biogeographical, and geographic ideas in the work derive in large part from Humboldt's friends, Karl Willdenow and

6. Of course, Newton's celebrated statement on originality was by no means original—see Merton 1993.

George Forster. Humboldt's accomplishments were in synthesizing these ideas, portraying them graphically using the concrete example of Chimborazo, and integrating them into a broader vision of science—a vision encompassing space, time, the physical and biotic elements of the earth, and human culture and perception. Even the synthesis, though unique, was not completely original; it built on ideas concerning science, form, and geography derived from two other towering figures of the time, the romantic poet (and scientist) Johann Wolfgang von Goethe and the Enlightenment philosopher Immanuel Kant. Details of the synthesis, and its specific application to Chimborazo, owe much to Humboldt's experience in the Andes and his discussions and tours with local naturalists, particularly Francisco José Caldas of Popayán and José Celestino Mutis of Bogotá. Humboldt combined all of these elements into a unique—and original—synthesis that set the stage for the rapid development of biogeography, ecology, and physical geography during the nineteenth century.

II

Humboldt's passion for botany was launched at age nineteen, when he met Karl Ludwig Willdenow during a visit to Berlin. Willdenow, only four years senior to Humboldt, had just published a flora of the Berlin region and was fascinated by the problems of plant distributions. He discussed plant geography and physiology with Humboldt, who was deeply influenced by Willdenow's ideas concerning the influence of climate on vegetation. Willdenow, in his 1792 Grundriss der Kräuterkunde,[7] articulated several of the central ideas Humboldt discussed in the Essay. Though Humboldt is often credited with these ideas, Willdenow recognized the dominant role of climate in governing plant geography and vegetation zonation, observed that latitudinal zonation

---

7. An English translation of Willdenow's 1792 work was published in Edinburgh as The Principles of Botany, and of Vegetable Physiology in 1805. Willdenow published a revised edition of Grundriss der Kräuterkunde in 1802; an English translation was published in Edinburgh in 1811. Both, of course, predate Humboldt's Essay and amply document the extent to which Humboldt's ideas on plant geography drew from Willdenow's. Willdenow's treatment of plant geography and ecology differs little between the editions, with one exception. In the 1802 (1811) edition, he revised his discussion of paleontology and earth history, incorporating Cuvier's views on global revolutions. The two English editions were translated by different, unnamed individuals with contrasting approaches to the same text.

is displaced northward in Europe relative to North America, noted that plants of polar regions grow on mountaintops at lower latitudes, demonstrated that plant diversity increases from pole to equator, and argued that vegetation is zoned latitudinally and not longitudinally. Humboldt later expanded these ideas, confirmed them with collections and observations, and devised effective graphical displays to convey the information to diverse audiences.

Early in 1789, Humboldt joined his older brother Wilhelm at the University at Göttingen, where he studied an array of subjects ranging from archeology and philology to physics and chemistry. At Göttingen, Humboldt met George Forster who, at eighteen, had set out with his father (Johann Reinhold Forster) aboard H.M.S. *Resolution* on Captain James Cook's second voyage of exploration around the world (1772–75). Forster and Humboldt became fast friends. In the spring of 1790, they traveled together down the Rhine, across Belgium to the French coast, and crossed the Channel to England. They returned by way of Paris, arriving there just in time for the first-anniversary celebrations of the fall of the Bastille. This trip, which was written up by Forster in a three-volume account,[8] had a tremendous influence on Humboldt. He sympathized with the French Revolution, was exposed to a variety of landscapes, languages, and cultures, and most importantly, spent countless hours listening to Forster's accounts of his South Sea voyages and discussing geography and science. Forster's father had recognized the role of climate in shaping plant form and distribution, and viewed vegetation pattern as the most important signature of the environment. George Forster adopted this view as well, expanding it to a unitary vision of geography, whereby all natural phenomena showed interlinked patterns in space. Forster's views grounded the next seven decades of Humboldt's thinking.[9]

After the trip with Forster and a boring interlude in business school in Hamburg, Humboldt traveled to Saxony for a stint at the Freiberg Academy of Mines—at the time the foremost academy of its kind in the world. Abraham Gottlob Werner directed an intensive and rigorous curriculum of geology, engineering, mathematics, chemistry, and surveying. In his daily trips to the pits and shafts of the local mines, Humboldt observed mosses, fungi,

8. George Forster, *Ansichten vom Niederrhein, von Brabant, Flandern, Holland, England, und Frankreich, im April, Mai und Junius 1790* (Berlin: Voss, 1791–94). Forster's account was reprinted several times during the nineteenth and twentieth centuries, but no English translations appear to have been published.

9. Humboldt generously and explicitly acknowledged George Forster's influence on his travels and thinking in volumes 1 and 2 of *Cosmos*.

algae, and vascular plants growing in seemingly unfavorable and lightless environments, and he studied them with what time he could spare. He spent the next five years as a mining inspector in Prussia, where he improved mining and extraction efficiency, opened a free training school for the unlettered miners (paid for out of his own pocket), and devised a number of mine-safety improvements, including portable breathing devices and safety lamps. At the same time, he completed and published his second scientific monograph,[10] *Florae Fribergensis* (1793). This work is principally concerned with the physiology of the subterranean plants, algae, and fungi that Humboldt found in the mines. However, it includes a discussion of the general approach to plant geography that Humboldt was developing in his mind, influenced by Willdenow and Forster, as well as by Immanuel Kant's essays on physical geography.

Humboldt's earliest scientific work was largely observational and synthetic. However, he soon demonstrated an aptitude for careful and systematic experimentation. Humboldt became interested in the recently discovered principle of galvanism, whereby muscles are stimulated by the application of electrical current. Between 1792 and 1797, he did extensive experiments on the subject, many on himself.[11] In the course of these studies, he came very close to stumbling on the principle of the battery, discovered a short time later by Volta. Humboldt described some 4,000 individual experiments, involving some 300 different kinds of animals and plants—mice and mimosas, leeches and tapeworms. He concluded that animals share the general potential for electrical stimulation (which is lacking in plants), and that the electrical stimulation is transmitted through nerves. The lasting significance of these studies, summarized in a two-volume monograph (1797), lies in Humboldt's rigorous, systematic application of reductionist, experimental science.

10. The first was a discussion of the basalts of the lower Rhine valley he observed during the excursion with Forster. Humboldt erroneously attributed them to waterborne processes, which greatly pleased Werner, his neptunist mentor. Humboldt later changed his views on basalt formation, recognizing their plutonic origins.

11. Many of these self-administered experiments were quite painful. He deliberately inflicted wounds and blisters on his own skin, and connected them with silver and zinc electrodes to verify his hypothesis that physiological electrical discharges ("animal electricity") could be conducted through the electrodes. He took advantage of a tooth extraction to perform similar experiments on his jaw, but the pain proved too severe to continue. It must be borne in mind that he performed these experiments at a time when bleeding and blistering were common in medical practice, and that he may have sought to combine experiment with remedy.

Humboldt's reductionist empiricism confronted its holistic, intuitive antipode when he met Johann Wolfgang von Goethe in 1795. Humboldt and Goethe shared interests in the sciences and visual arts, and it is a credit to their powerful and nimble intellects that instead of colliding they became close friends and mutual admirers. Goethe had already established himself as a premier writer with *The Sorrows of Young Werther* and numerous poems, and he had for some time also been preoccupied with the morphology of plants and the physics of color. He had published his *Metamorphose der Pflanzen* (Metamorphosis of Plants) five years earlier, in which he made a series of arguments that were to influence Humboldt's perception of nature and vegetation. Goethe's concept of organic morphology (a word he coined) was central to his thinking. He emphasized the features that all plants held in common (rather than those which differentiated them, which Linnaeus and his followers focused on), and viewed plant genera, species, and individuals as variations of archetypal forms. Goethe also observed that the outer forms of plants and animals reflected their environments, noting, for instance, that the submersed leaves of *Ranunculus aquatilis* differed morphologically from the aerial leaves, and that individuals of the same species varied morphologically when grown in different habitats. Finally, he viewed form as an index to underlying processes—by understanding surface features, fundamental insights would emerge. These notions were to play important roles in Humboldt's ecological thinking.

Through much of the 1790s, Humboldt dreamed of traveling to far parts of the world to make observations and measurements, inspired by Willdenow's ideas on plant geography, Forster's accounts of his adventures in the South Pacific, Kant's vision of a science of physical geography, and Goethe's notions of plant and landscape form. His mother's death in late 1796 left him with a considerable fortune—enough to quit his government post and finance a voyage on his own. A succession of planned trips were diverted or aborted as Europe sank into the chaos of the Napoleonic Wars. In 1798 he traveled to Paris to visit his brother Wilhelm and pursue opportunities there. He met the great French explorer Louis Antoine de Bougainville, who at 70 was planning a five-year global scientific expedition. Bougainville invited Humboldt to join the expedition, which Humboldt immediately accepted. But the plans unraveled, first when the French government replaced Bougainville with a younger captain, Nicolas Baudin, and then, less than two weeks before departure, by indefinite postponement of the voyage. France needed to finance yet another war, this one with Austria.

In the preparations for the Bougainville/Baudin expedition, Humboldt had befriended Aimé Bonpland, a talented French botanist who shared Humboldt's desire to travel. After an aborted trip to Egypt, Humboldt and Bonpland walked from Marseilles to Madrid, collecting plants to send to Willdenow and making detailed observations and measurements of the landscape. They were the first to demonstrate that the Spanish "plain" was an elevated plateau, which explained its unusually continental climate.

Arriving in Madrid in February 1799, Humboldt, through a serendipitous chain of personal connections, was able to obtain an audience with King Carlos IV (Humboldt had once met the brother of the Saxon ambassador, who knew the prime minister, who was the paramour of the queen, who generally could get the king to do her bidding). Humboldt proposed to self-finance a scientific expedition to the Spanish colonies of the New World, suggesting that his voyage might yield discoveries of useful plants, mineral deposits, and other treasures. The king provided him with an open-ended passport— unprecedented for a foreigner, and a Protestant one at that. On 5 June 1799, Humboldt and Bonpland were aboard the Pizarro, bound for South America.

III

Humboldt's "American travels . . . have as completely changed the basis of physical sciences as the revolution which took place in France about the same time has changed the social condition of that land."[12]

" . . . we have been running around like mad things and in the first three days were unable to make any clear observations because we tended to let one thing drop in order to grasp the next thing that offered itself. Bonpland says he will really go mad if there is not soon an end to the wonders."[13]

Oceangoing voyages were hazardous in 1799, and the voyage of the Pizarro was made more so by Spain's alliance at the time with France, which was at war with England. British warships patrolled the coasts of Spain and its possessions, eager to seize any Spanish merchant ship or packet vessel that came in their path. The Pizarro slipped out of the harbor at Corunna suc-

12. Louis Agassiz, Eulogy by Professor Agassiz upon Baron von Humboldt, in Agassiz 1859, 646.

13. Letter from Alexander von Humboldt to Wilhelm von Humboldt, Cumaná, dated 16 July 1799, in Meyer-Abich 1969, 49.

cessfully but had a brief scare a few days out when it spotted a distant British flotilla while heading toward the Canary Islands. The ship's captain was under royal orders to stop at Tenerife so that Humboldt and Bonpland could explore the volcanic landscape. The ship narrowly escaped capture by four British warships that materialized out of the fog near the harbor at Santa Cruz. Humboldt and Bonpland debarked at Santa Cruz and were joined by the French vice-consul for a climb to the 3,715-m peak of Pico de Teide, the highest mountain on Tenerife. Although Humboldt had scaled passes in the Alps and Pyrenees, this was his first experience on a tropical mountain. He noted the vegetation zones and temperature changes as the group scaled the peak. The sulfuric acid gassing from the peak's crater burned holes in their clothes and notebooks as the party collected rocks and gases and measured temperature, barometric pressure, and the blueness of the sky.[14] The volcanic rocks, conducting heat from the magma chamber below, scorched their feet while near-freezing wind gusts chilled their hands and faces as they explored the summit.[15]

The *Pizarro* anchored in the port of Cumaná on the coast of Venezuela in mid-July. Although Humboldt and Bonpland originally planned to continue on to Havana, fatigue from three weeks at sea, a typhoid outbreak aboard ship, and the exotic world they entered in South America convinced them to stay. Virtually everything they experienced was new—parrots, palm trees, monkeys, brightly colored crabs and flowers, armadillos, crocodiles, native peoples. It took them some weeks to prepare for systematic scientific work, between the continual barrage of new sights, smells, and sounds, and the stream of local visitors, eager to examine all the scientific instruments and view miniature wonders through the microscope Humboldt had brought along.

Humboldt and Bonpland spent sixteen months in Venezuela. After four months making botanical observations and meteorological measurements

14. The instruments Humboldt used in his travels are described in an accompanying essay.

15. Humboldt's account of the visit to Tenerife, the ascent of the Pico de Teide, and his digressions on the landscape, geology, vegetation, agriculture, meteorology, and cultural history of Tenerife occupy some 214 pages of the Williams translation of his *Personal Narrative*. I have found no better introduction to the breadth and comprehensiveness of Humboldt's scientific vision at the time of his voyage. Humboldt's passion for the study of nature, his lyrical descriptions of landscapes, and his sympathy for the enslaved and downtrodden are all apparent in this passage.

in Cumaná and exploring the surrounding mountains, they took a boat to Caracas. They climbed the Silla, the tallest peak above Caracas, where Humboldt noted a number of plant species typical of the Andes, several hundred kilometers to the southwest.[16]

In February 1800, the explorers left Caracas for the Venezuelan interior, traversing the valley of Lake Valencia and setting out across the llanos towards the Orinoco River. They spent a few days in Calobozo, where they captured electric eels and performed an array of experiments on them, getting shocked in the process. They traveled on to the Río Apure, a tributary of the Orinoco. It was here that they embarked on an ambitious expedition to determine whether Fray Antonio Caulin (a Spanish missionary who traveled extensively in the region in the mid-1700s),[17] was correct in his controversial claim that the upper Orinoco split, with a trunk stream, the Casiquiare, flowing into the Río Negro in the Amazon drainage. Humboldt and Bonpland, together with Don Nicolas Sotto, a relative of the provincial governor, Padre Bernardo Zea, a local missionary, and nine native guides and paddlers, set off down the Río Apure, up the Orinoco, past a 65-km series of rapids, and up a tributary stream, the Río Atabapo. They carried the dugout canoe and gear overland four days to reach the Río Negro. Descending the Río Negro, they found the Casiquiare. By this time the party had accumulated an array of parrots, toucans, manakins, and several kinds of monkeys, all in cages which they crammed aboard the dugout. Biting insects, which had plagued the party throughout the trip, grew steadily more abundant, diverse, and aggressive as they ascended the Casiquiare. This did not deter Humboldt from making exacting astronomical measurements whenever the night skies were clear enough, nor Bonpland from collecting, describing, pressing, and mounting hundreds of plant specimens. Supplies were low, and the party supplemented their meager rations with ants. Eventually they reached the Orinoco, confirming Caulin's account, and headed downstream and back across the llanos.

16. Humboldt's description of the ascent of the Silla also makes for fine reading. It is much briefer (55 pages in Williams's translation) than the Tenerife passage, and includes a number of botanical observations and insights.

17. Caulin's *Historia corográfica, natural y evangélica de la Nueva Andalucía, Provincias de Cumaná, Nueva Barcelona, Guayana y vertientes del río Orinoco*, published in Madrid in 1779, summarized the geography, natural history, native peoples, and philology of the Orinoco and other regions of modern Venezuela and Guyana. Humboldt consulted the volume closely.

The expedition of Humboldt and Bonpland followed a serendipitous course. Late in 1800, they sailed to Cuba aboard a cargo ship, planning to sojourn in Havana, curate specimens and organize notes, and then proceed to the United States, where they would traverse the Great Lakes and Ohio and Mississippi Valleys before embarking for Mexico and the South Pacific. But in Havana, they saw a newspaper account that Captain Baudin's voyage of exploration had finally been authorized and that he was expected in Lima in a year's time. In March 1801, they took passage on a ship bound for Cartagena on the northeast coast of South America, with plans to cross the isthmus of Panama to the Pacific, and then take another ship to Lima. They arrived in Cartagena just after the trade winds ceased for the season, and so they decided instead to trek across South America to meet Baudin.[18]

After six weeks by canoe up the Magdalena River to Honda, followed by a 9,000-foot climb along an "indescribably bad" road—an Inca trail unimproved by three centuries of colonial Spanish use—Humboldt and Bonpland reached Santa Fe de Bogotá. They became instant celebrities in a remote city that saw few foreigners, and were hosted by Don José Celestino Mutis, South America's premier botanist and naturalist. Humboldt spent two months in Bogotá, exploring the surrounding territory and examining Mutis's considerable herbarium and botanical library while Bonpland recovered from malaria.[19]

Leaving Bogotá in September 1801, Humboldt and Bonpland set out on foot for Quito, crossing mountains and valleys as the rainy season commenced, trekking across the frigid páramos of Pasto, traversing rainforests, ravines, and volcanoes, sleeping under tents improvised from imbricated Heliconia leaves. They reached Quito in January and stayed there for six months, using it as a base for explorations of the surrounding Andes. Humboldt was joined en route to Quito by Francisco José Caldas, a self-taught naturalist and astronomer from Popayán, whose ingenuity and knowledge impressed Humboldt. Caldas spent several weeks with the party. Although

18. The *Personal Narrative* ends shortly after Humboldt and Bonpland's arrival in Cartagena. Humboldt prepared an additional volume detailing the adventures in the Andes but destroyed it shortly before publication, for reasons that remain obscure.

19. Humboldt seemed to have a remarkable resistance to disease. He was in good health for nearly the entire voyage, despite repeated exposure to the elements, strenuous activities, and bites from the various parasites and insects. Bonpland, in contrast, suffered from a variety of tropical diseases and other misfortunes while in South America.

he had no formal training in botany at the time, he knew vegetation and was interested in the patterns of vegetation in relation to altitude and climate.[20] He was also intimately familiar with the geography of the region, having climbed many of the peaks to measure their altitude. Humboldt and Bonpland taught Caldas the formal elements of botany, and Humboldt briefed him extensively on scientific developments and current knowledge in Europe.[21]

Caldas, like Forster, Willdenow, and Goethe, is an important figure in the development of Humboldt's *Essay*. Humboldt's ideas concerning plant geography were well developed by the time he met Caldas. However, they spent several weeks together, during which Caldas exchanged his knowledge of the vegetation, climate, and geography of the central Andes for Humboldt's conceptual ideas on plant geography and his (and Bonpland's) knowledge of plant classification. These discussions, and the excursions they took

20. Humboldt has been charged, most recently by Margarita Serje (2004), with stealing the central concepts of the *Essay* from Caldas. Serje asserts that the concept of plant geography was "first and exclusively invented by Caldas. Humboldt took advantage of these inventions, without acknowledging or citing this fact in his works." These charges are completely without merit and can be refuted by no more than a cursory look at Humboldt's botanical work (as well as that of Forster and Willdenow) before he ever reached South America. Cañizares-Esguerra (2005) provides an excellent account of the issue (see also Appel 1994). It is true that Caldas was invaluable in helping Humboldt understand the geography of the Andes (as was Mutis), and that Humboldt benefited greatly from discussions with Caldas on the vegetation and climate of the region. It is also true that European and North American scholars have, until recently, neglected Caldas's intellectual accomplishments and his contributions to Humboldt's efforts. Similar charges leveled by Serje that Humboldt adopted Caldas's boiling-water method for measuring altitude are also false. Humboldt was impressed by the method, but he had already invested three years in barometric measurements of altitudes in South America and continued to rely on barometry for the rest of the voyage. Humboldt used many of Caldas's altitudinal measurements, with Caldas's approval. Appel (1994) provides a thorough account of the relationship between Humboldt and Caldas.

21. Caldas had limited access to scientific works, mostly outdated ones, in his remote home town of Popoyán. He independently discovered the principle of hypsometry (the contingency of the boiling temperature of water on atmospheric pressure) in early 1801 and, until he met Humboldt, had no way of knowing whether the principle was already known in Europe. It was, having been discovered by Fahrenheit some 67 years before. Saussure had already conceived that the boiling-point temperature could be used as a metric for altitude. However, Caldas's calibrations and formulae were much more accurate than Saussure's; Caldas recognized that the air-pressure/elevation relationship was not linear and incorporated this into his calculations.

together in the region, may well have crystallized Humboldt's decision to use the Andes as the ideal region to illustrate his ideas (e.g., in the Chimborazo profile). Humboldt certainly stimulated Caldas's thinking on plant geography; after Humboldt's departure, Caldas drafted profiles of the Andes similar to Humboldt's profile, with specific plant species drawn into their appropriate elevational and topographic settings. He also prepared a monograph on the geography of plant cultivation in the region.[22] Caldas evidently also prepared other works and charts on plant geography, but all appear to have been lost in the turmoil of revolution, civil war, and Caldas's arrest and execution after the Spanish reconquest.

During an extended stay in Quito, Humboldt and Bonpland befriended Carlos Montufár, the son of the provincial governor, who accompanied them in many of their excursions through the Andes, and in fact accompanied them for the rest of the voyage, returning to France with them. Montufár's father seems to have sponsored his trip in hopes of his developing some focus and receiving a broad education in Europe. Montufár was not terribly engaged by science or natural history (his ultimate career was military), but he was evidently a willing companion and field assistant, up to the physical challenges of the expedition. On 23 June 1802, Humboldt, Bonpland, Montufár, and a native guide climbed up the slopes of Chimborazo, a volcano considered at the time to be the world's tallest mountain. The party ascended to 6,327 m, a new climbing record not matched until the botanist Joseph Hooker surpassed it by a few meters in the Himalayas in 1849.[23] Humboldt and his companions stopped 450 m short of the Chimborazo summit, their passage blocked by a deep ravine.

Throughout the Andes explorations, Humboldt maintained his hopes of intercepting Baudin's ships in Lima and sailing to the South Pacific. While in Quito, he learned that Baudin had changed his route, rounding Cape

22. Reprinted as an appendix in Appel 1994.

23. The maximum elevation Hooker attained is unclear, though he spent several weeks in September 1849 exploring territory well above 5,000 m. Hooker and his party climbed the slopes of Donkia to "between 19,000 and 20,000 feet" (6,234–6,562 m), and noted that he observed a "Brocken spectre" at 19,300 feet (6,332 m). It is likely that unrecorded Tibetans and Nepalese had already attained higher elevations; Hooker noted that Tibetans routinely crossed passes exceeding 18,500 feet (6,070 m). The first successful ascent to the summit of Chimborazo was not accomplished until 1880, by Edward Whymper of Matterhorn fame. Not one to do anything halfway, Whymper led two different groups to the top using two different routes.

Horn instead of the Straits of Magellan, and that he would never reach South America. Humboldt and his party left Quito and trekked south to Peru and the Amazon headwaters region. Humboldt spent several weeks there, investigating Inca ruins and artifacts and researching the cinchona tree (source of quinine bark), before heading southwest, crossing the Andes near the equator. After sixteen months in the Andean highlands, Humboldt descended to sea level and reached Lima in late October 1802. He and his party made their way up the Pacific coast by ship to Guayaquil, and then boarded a ship for Acapulco, arriving there in March 1803.

Humboldt, Bonpland, and Montufár spent nearly a year in Mexico, primarily in Mexico City. Humboldt spent much of the time in offices and libraries, compiling the information and data he later published in his groundbreaking regional monograph, *Political Essay on the Kingdom of New Spain*, which summarized the geography, economics, and politics of the region, and which featured novel statistical graphics still in use today.[24] Humboldt took time for a few excursions, particularly to visit the various volcanoes of the Valley of Mexico, the silver mines of Guanajuato, and the pyramids and other archeological ruins of the region. As always, he took exacting astronomical and meteorological measurements and made botanical and geological observations on a daily basis. En route from Acapulco to Mexico City, and thence to the port of Veracruz, Humboldt made detailed observations of the bedrock geology, which he later published in pioneering longitudinal profiles.[25] The party sailed from Veracruz for Havana in March 1804, where they spent a few

24. These graphics included subdivided histograms and superimposed proportional squares. In the *New Spain* volume, he adopted statistical graphics previously used to portray physical properties and applied them to social and economic data. Humboldt's graphics were inspired by William Playfair and others; the late eighteenth and early nineteenth century represent the dawning era of modern statistical graphics. Humboldt's most original contributions were his development of geographic isolines (e.g., his 1817 isotherms) and his novel use of topographic profiles (as in the *Tableau physique*).

25. The profiles were analogous to the Chimborazo profile, except that they emphasized the belowground geological features rather than the aboveground vegetational and floristic features. Humboldt summarized his geological and stratigraphical views in his 1823 *Geognostic Essay on the Rock Formations of the Two Hemispheres*. He generally receives less credit than his English contemporary, William Smith, for developing the principles of stratigraphy and correlation, largely because of his reliance on mineralogical characters and relative neglect of fossils (which Smith recognized as a primary source of chronological and stratigraphic information).

weeks preparing the plant, animal, and mineral collections deposited some three years before for transport to Europe.

Humboldt and his companions detoured to the United States on the way back to Europe, spending a few weeks in Philadelphia, at the time the intellectual center of the country, visiting with the various naturalists, scientists, and artists in residence. They journeyed to Washington, D.C., at the invitation of President Thomas Jefferson. Jefferson was an avid naturalist and scientist, and by all accounts he and Humboldt quickly developed a warm friendship (they corresponded until Jefferson's death). After receiving a *laisser-passer* from the British Consul in Philadelphia (France and Britain were still at war, and Britain was blockading French ports and seizing French ships), Humboldt, Bonpland, and Montufár embarked aboard a French ship bound for Bordeaux.

Humboldt arrived in Paris in August 1804. Aside from trips to Prussia and excursions to England, Italy, and the Alps, he spent the next twenty-two years there, giving lectures, doing experiments, preparing seemingly countless monographs and books, all stemming directly or indirectly from the South American tour, and spending the rest of his considerable fortune on assistants, artists, and publication costs. He was under steady pressure to return to his native Prussia, and as a foreigner in the Napoleonic era was under constant police surveillance and censorship. Humboldt enjoyed the Parisian lifestyle, to be sure, but his primary reason for staying was the intellectual, artistic, and technical expertise there. Paris was, at the time, the intellectual capital of Europe, with leading botanists (Jussieu, Desfontaines), zoologists (Lamarck, Geoffroy Saint-Hilaire), geologists (Cuvier, Brongniart), physicists (Laplace, Arago, Biot, Delambre) and chemists (Gay-Lussac, Berthollet, Vauquelin). In contrast to the centralized French government and culture, which concentrated the nation's talents in the capital city, Germany was still a diffuse and diverse assortment of kingdoms, principalities, and city-states. Only a fraction of Prussia's considerable scientific and technical talent lived in Berlin; others were dispersed from Königsberg (now Kaliningrad in Russia) to Breslau (now Wroclaw in southwest Poland) to Frankfurt-am-Main (now in western Germany)—all before railroad or telegraph.

The *Essay* was the first major published work produced from the voyage, appearing in 1807.[26] Its frontispiece lists it as the first volume of the first part

---

26. Many copies were printed with an 1805 date, and it is frequently cited as an 1805 work. Although there is still some confusion surrounding the actual date of publica-

of a multivolume series collectively entitled, *Voyage of Humboldt and Bonpland*. The first part was intended to cover "general physics and an historical account of the voyage." Just how many parts and volumes Humboldt planned for the series is not entirely clear; the writing project evolved over the next two decades, and Humboldt had to modify his plans as his financial resources dwindled. Ultimately, the series included summaries of astronomical and barometric observations (2 volumes), botanical monographs and synopses (20), zoological and anatomical observations (2), political essays on the island of Cuba (3 octavo) and New Spain (3 quarto), geographic and pictorial atlases (2), and the personal narrative (13 octavo volumes). Several of these volumes were printed in both folio and quarto (or quarto and octavo) versions, and some were reprinted as separates or extracts. Humboldt intended one or more volumes on geology but never completed them (though he did publish an extended essay comparing the stratigraphy of Europe and South America in 1823). Most of the original volumes are in French, except many of the botanical monographs, which are in Latin.[27]

Humboldt later listed the *Essay* as the sole volume in part 5 of the series. Its initial appearance as the first volume in the series probably reflects its readiness when he completed the voyage (he had been pondering the topic for nearly two decades and wrote much of the text in South America). However, its primacy undoubtedly also stems from the primacy of plant geography in Humboldt's overall vision of the world, whereby vegetation is both the most obvious surface manifestation of climate and the determinant of many other natural and human features.

IV

"To the celebrated Humboldt . . . we are indebted for the most valuable writings on vegetable geography, which have first given it the true character of a science."[28]

---

tion, it appears that no copies were in circulation before 1807, though Humboldt had finished the work in 1805 and had presented it as a lecture to the Paris Classe des sciences physiques et mathématiques in January of that year. See Stearn 1960 and Drouin 1997.

27. Humboldt's 1817 *De Distributione Geographica Plantarum* was published as the prolegomena to the *Nova Genera et Species Plantarum* and later as a separate octavo volume. This work was an extension and elaboration of the *Essay* and has never been translated into English.

28. William Jackson Hooker 1837, 237.

"The history of the products of the soil is closely interwoven with the fate of mankind and its emotions, ideas, and actions. The realm of nature borders the domain of every science, and it is impossible to review the former without examining the latter."[29]

Humboldt's Essay volume is organized into four parts: a brief "Preface," an "Essay on the Geography of Plants," the much longer "Physical Tableau of the Equatorial Regions," and the color plate, the *Tableau physique des Andes et Pays voisins.*[30] Humboldt's "Preface" makes clear from the start that this is an ambitious work. In his second paragraph, he states boldly that this work "consider[s] together all the physical phenomena that one can observe on the surface of the earth as well as in the surrounding atmosphere" (61). Humboldt intends to tell us about more than plant geography; he is going to show how diverse phenomena of the world can be unified and reduced to a small set of interconnected patterns. This unification is not fully revealed, however, until the reader reaches the *Tableau physique*, the Chimborazo profile (described in section V).

In the "Essay," Humboldt exhorts botanists to go beyond collecting, describing, and classifying plant specimens. He never questions the usefulness of these tasks—indeed he devoted countless hours to them in his lifetime, and he spent a considerable sum supporting Kunth's taxonomic work on the plant collections of the American voyage. But he appeals to the audience to go beyond classification, to focus on the geography of plants, a science he describes as "an essential part of general physics." This seems like a curious statement today; "general physics" calls up images of ironclad equations and

29. George Forster, date unknown (published 1843). Quoted in Erwin H. Ackerknacht 1955.

30. In Humboldt's French text, the operative noun is *Tableau*, and in the German version it is *Naturgemälde.* In both languages the nouns have multiple meanings (e.g., "description"), but the most direct translation, and the meaning Humboldt undoubtedly had in mind, was "picture" or "painting." For the current translation, we decided to use "tableau" in its primary English usage of "a vivid or graphic description." In the *Essay*, the Chimborazo profile was literally the picture, and the text was an extended essay to accompany and explain the picture. This reflects Humboldt's fascination with the visual arts, and his long-term effort to fuse art and science, as manifested in the second volume of *Cosmos* and several essays in *Views of Nature.* Stephen Jay Gould, Anne Godlewska, Aaron Sachs, Edmunds Bunkśe, Michael Dettelbach, and others discuss Humboldt's humanistic vision of art and science as complementary routes to understanding the world.

universal laws, and seems far removed from notions of either plants or geography. Humboldt uses physics in its original, elemental sense: knowledge of the material world based on observation and experimentation (i.e., science).

The "Essay" appears rambling and discursive to the modern reader, unusual for a scientific work. Humboldt was capable of focused, rigorous scientific prose, as can be seen in much of the "Physical Tableau" text. He adopts a less formal, often lyrical, style in the "Essay," which is as much an aesthetic statement as a scientific thesis. His aim is not to inform the reader about the *findings* of plant geography; those will come later in the "Physical Tableau." Rather, he wants to convey a vision of what plant geography is, or can be, and why it matters intellectually, practically, and aesthetically.

Humboldt offers a series of scientific theses in the "Essay," interlaced with prose relating them to a variety of humanistic concerns, ranging from classical history to human sustenance to artistic perceptions. He first defines plant geography as the study of plant distributions in relation to climate, though he says little about climate in the rest of the "Essay." In his first extended discussion, Humboldt classifies plants into two general categories of intraspecific sociability: those that occur primarily as dispersed, solitary individuals and those that are "socially organized." Of the latter, we would recognize most today as clonal and the others as tending to grow in monospecific stands. Humboldt delineates any number of ecological concepts that remain widely used two centuries later: plant community, ecological diversity, vegetational inertia (resistance to invasion or other change), and the mutual influences of vegetation and environment.

Humboldt goes on to acknowledge and discuss the role of history in influencing plant distributions. His treatment of history here is equivocal.[31]

31. Humboldt was generally averse to historical explanations for biogeographical patterns early in his career. In the *Personal Narrative*, he repeatedly discusses phenomena, both biogeographical and geological, that require historical explanation. In each case, he acknowledges the necessity of history and then retreats from formulating or evaluating any specific historical explanations. In fact, he frequently asserts that the problems are insoluble and that historical hypotheses are untestable. For example, in discussing disjunct populations of Andean plants on the Silla of Caracas, he observes that there is no continuous corridor of suitable, high-elevation habitat between the two regions. He then states that "[t]he more we study the distribution of organized beings on the globe, the more we are inclined, if not to abandon the ideas of migration, at least to consider them as being hypotheses not entirely satisfactory . . . " and goes on to assert that "these problems cannot be solved. . . . [T]he task of the philosopher is fulfilled, when he has indicated the laws, according to which nature has distributed

On the one hand, he recognizes the preponderance of evidence for biogeographic and environmental change in earth history. Neither modern plant distributions nor fossil occurrences can be explained without invoking history. At one point, he even makes a concession to the possibility of organic evolution.[32] On the other hand, he is clearly challenged by the magnitude of climate change required to explain fossil evidence for tropical plants and animals in Europe, and he leaves open the possibility that the fossils were transported to high latitudes. He describes the kinds of mechanisms that might drive such climatic change—changes in planetary dynamics relative to the sun, cooling of earth's core, variability in solar radiation. All of these mechanisms would be the focus of debate in the nineteenth century as the science of historical geology developed. Humboldt leaves the question open, appropriately in view of the dearth of an organized geological record, but highlights the intellectual riches to be gathered by considering plant geography together with earth history.

An extensive discussion is devoted to human utilization of plants and the role of humans in the migration of useful plants. Humboldt strives for generalizations (rooted in Forster's writings) about the influences of climate and vegetation in societal organization and resource utilization by humans. As in the case of earth history, he is agnostic about specific historical explanations

---

vegetable forms" (Williams translation, 3:494–95). Humboldt was clearly aware of geologic history and its potential for influencing biogeographic distributions, but was unwilling to take advantage of it in providing explanations. A coherent science of earth history was lacking at the time Humboldt wrote the *Essay* and *Personal Narrative*, though Cuvier, Hutton, Smith, and others were developing the underpinnings. It remained for Charles Lyell, followed by Edward Forbes and Charles Darwin, to integrate biogeography with historical geology. Humboldt embraced this integration in *Cosmos*, quoting Forbes's writings at length in volume 1 (footnotes on 348–49 and 352–53).

32. Humboldt states at the bottom of page 68 that practitioners of plant geography can identify "primitive" plant forms, and raises the question "whether the diversity of species can be considered to be an effect of the degeneration that over time transformed accidental varieties into permanent ones." He is acknowledging the thinking of Lamarck and other contemporaries on the plasticity of species and the possibility of organic evolution. Later in the "Essay" he makes reference to Cuvier's Egyptian ibis mummy (obtained during Napoleon's invasion of Egypt, initially described by Cuvier in an 1804 paper and discussed again in his 1826 *Discours sur les révolutions de la surface du globe*), adopting Cuvier's interpretation that evolution had not occurred in historic time. Humboldt was ambivalent about evolution throughout his lifetime; he died a few months before publication of Darwin's *Origin of Species*, which might have settled the question for him.

for specific plant distributions, acknowledging only that modern distributions of cultivars have historical roots. He uses accounts from classical history to demonstrate that humans have moved plants—e.g., the proliferation of cherry orchards throughout the Roman Empire in Europe within a few decades of Rome's conquest of Pontus (modern coastal Turkey)—but is reluctant to identify specific places of origination. Humboldt clearly recognized the problems of historical inference, and the "Essay" predates the development of formal methods of historical science by Cuvier, Lyell, Murchison, and others. Humboldt identified important questions concerning domestication, but answers came only decades later with the work of Charles Darwin and Alphonse de Candolle.[33] Many did not emerge until the advent of genetic analysis and radiocarbon-dating in the twentieth century.

Humboldt concludes this discussion with a coda (page 72) noting the importance of plant geography in human history, both political and intellectual. He then moves to the aesthetics of vegetation. After some rhapsodic prose on the central importance of vegetation in our sensory perception of landscapes, he makes a novel proposal that plants be classified superficially, in terms of their gross form.[34] Rather than abandoning Linnaean classification, he advocates an alternative, complementary classification of plants. Humboldt's discussion centers on aesthetic perception, but he is clearly groping towards an essentially ecological concept, rooted in Goethe's notion of form as a key to function. The geographic patterns in occurrence and dominance of different plant-forms must surely indicate something fundamental about plants and vegetation in the context of their environment. That Humboldt's intuition was correct is borne out by the continuing ecological and biogeographic emphasis on plant physiognomy, both in understanding organismal adaptation to environment and in the influence of plant form on ecosystem properties.[35]

33. Darwin's 1868 *Variation of Animals and Plants Under Domestication* and Candolle's 1884 *Origin of Cultivated Plants* are regarded as the foundation works on plant domestication. They clearly owe much to Humboldt.

34. Humboldt lists fifteen types of plant-form. These are not intended to be inclusive or exhaustive. His essay "Ideas for a Physiognomy of Plants" in *Views of Nature* (1808) provides a fuller account of plant-form, with several types not included in the "Essay." He notes that his emphasis is on the tropics and that many additional plant-forms are yet to be described.

35. The physiognomy of plants, both individually and collectively (in vegetation), has a rich history in ecology. The pioneering plant ecologists of the late nineteenth and early twentieth centuries (Warming, Clements, Raunkiaer, Schimper, Cowles)

Although written in a rambling style, Humboldt's "Essay" amounts to a manifesto calling for the primacy of vegetation in scientific, practical, and aesthetic considerations of the landscape. Vegetation is at the heart of a "general physics" of the earth's surface. Vegetation cloaks that surface and governs exchange of energy and materials between the atmosphere and the surface. Vegetation provides plant resources that people utilize, and its physical structure and dynamics determine the capacity of the land for human activities ranging from hunting and gathering to clearance and cultivation. And vegetation dictates the aesthetic characteristics of the landscape. In Humboldt's vision of a unified science of the earth, the plants inhabiting the surface provide the foundation for a general description of the earth, serve as the focal point for intellectual, aesthetic, and practical considerations by people, provide important clues on the history of the earth, and represent explanatory challenges in terms of the physical and chemical properties of the atmosphere and soil. Humboldt's "Essay" sets the stage for the *Tableau physique* and accompanying text, where he makes his initial attempt at a comprehensive description and explanation of the earth's surface.

V

" . . . *statistical projections, which speak to the senses without fatiguing the mind, possess the advantage of fixing the attention on a great number of important facts.*"[36]

"*Before Humboldt we had no graphic representation of complex natural phenomena which made them easily comprehensible, even to minds of moderate cultivation.*"[37]

---

were deeply concerned with plant form and vegetation structure. This interest continued through twentieth-century plant ecology, and vegetation physiognomy is receiving intensive study in ecosystem ecology and biogeochemistry at local to global scales. The physical structure of the dominant plants on the landscape mediate the exchange of energy, carbon dioxide, and water between the earth's surface and the atmosphere, and can influence climate and weather at regional scales. Plant form also affects hydrology and chemistry of surface waters and groundwater, as well as the structure and chemistry of soils. Humboldt anticipated many of these influences, particularly vegetation/atmosphere interactions, in the *Personal Narrative* and in *Views of Nature*.

36. Alexander von Humboldt, *Political Essay on the Kingdom of New Spain* (1811), cxxxiii.

37. Louis Agassiz, Eulogy by Professor Agassiz upon Baron von Humboldt, in Agassiz 1859, 647.

The centerpiece of Humboldt's *Essay* is the color profile of Chimborazo, entitled *Geography of Equatorial Plants: Physical Tableau of the Andes and the Neighboring Countries*. The *Tableau physique* is a longitudinal slice of equatorial South America, viewed from the south (west is to the left). It is highly vertically exaggerated and is entirely schematic, emphasizing Chimborazo and its neighbor, Cotopaxi, as idealizations of the Andes. The notch between Chimborazo and Cotopaxi is halfway across the figure, and the lower limit of permanent snow is slightly more than halfway up the vertical axis. Thus, the entire botanical portion of the figure is in the lower half, and the upper half is devoted to the sky and the small, unvegetated cones of the Chimborazo and Cotopaxi summits. Various cloud layers are portrayed in the atmospheric profile, and the sky grades from white at sea-level upward to pale blue and ultimately rich, dark blue. The left-hand body of Chimborazo is cloaked in green vegetation, with palms discernible among the trees on the lower slopes, and trees of various forms arrayed along the slopes until they yield to short-statured plants above the cloud forests. Eventually the continuous vegetation cover is broken up by patches of exposed rock, which give way to permanent snow.

The right-hand portion of the land profile is a white background filled in with the names of individual plant species and vegetation zones. These labels portray the characteristic locations of plant species and vegetation types along the elevational gradient from sea-level to snowline and, to a lesser extent, laterally along the east slope of the Andes. A deep notch in the eastern portion of the profile portrays the interior valleys and lower mountains of the eastern Andes, which Humboldt explored extensively.

The vertical axis of the figure is to scale, and spans some 10,000 meters total (from more than 1,000 m below sea level to more than 9,000 m above). The vegetation profile rests on a brown foundation, representing the solid continent below sea level. Some text is visible in the brown portion, representing the subterranean "vegetation" (fungi, algae, mosses) that fascinated Humboldt since his days in the Freiberg mines. Marine plants (*Fuci, Ulva*) are labeled beneath the ocean waters. The sky is labeled with a number of altitudinal reference points—the summits of Vesuvius, Pico de Teide, Mont-Blanc, and Popocatépetl, the city of Quito, the point to which Humboldt and his party ascended Chimborazo in 1802, and the point to which Gay-Lussac ascended in a balloon above Paris in 1804.

The figure is flanked on each side by text-columns, eleven on the left and nine on the right. Viewed as a whole, the ensemble resembles a medieval triptych; the two flanking panels supplement the central illustration. Two columns on each side provide the vertical scale, in meters and in toises (fathoms) relative to sea level. The remaining sixteen provide measurements, descriptions, or discussions of various phenomena as they vary (or do not vary) systematically with altitude. These sixteen columns do not seem to be in any particular order. Each has a corresponding text-section in the "Physical Tableau" portion of the *Essay*, which discusses them in a logical order, from physical and atmospheric properties to land-surface phenomena.

The *Tableau physique* ensemble is a marvel of graphical portrayal, with a remarkable amount of information presented elegantly and economically.[38] Particular care was given to construction of the central figure. The colored portions convey a vivid sense of changes in the land surface (specifically vegetation and snow cover) and atmospheric properties (blueness, cloud pattern) with altitude. It also provides reference points for the European reader to judge. For instance, the reader can readily observe that Mont-Blanc, the highest summit in Europe, is at the lower limit of permanent snow in the Andes, and that the height of Mount Vesuvius is lower than the relief between the interior Andean valleys and the Andean summits.

The black-and-white portion of the central panel provides a detailed account of the plant geography of the Andes. This section, although appearing cluttered at first, is rich in information. This is no haphazard jumble of plant names spewed onto the mountain profile. Humboldt took special care to show where particular plant species, genera, families, and communities occur along the Andean slopes. The sloping text for vegetation zones (*régions*) and some plant species was intended to convey the sense that they spanned a wide elevational range. Horizontal text often indicates a discrete elevational threshold (e.g., the upper limits for *Cinchona* at 2,900 m and for trees at 3,400 m). Variations within the Andes are shown by the changes in vegetation zones from left to right at equivalent elevations, and to a lesser extent by the left-to-right pattern of plant species and genera. These are not always longitudinal; *Rhizophora*, a coastal mangrove tree, is shown at sea level but on the left-hand portion of the diagram in the continental interior.

The flanking black-and-white panels in the triptych are also packed with information. The columns, translated in a separate table, include (from left to right on Humboldt's figure):

---

38. Sylvie Romanowski's accompanying essay provides a more detailed analysis of the aesthetic and informational qualities of the *Tableau physique*, placing it in context of the development of scientific illustration and eighteenth century science.

1. Light refraction at 1,000-meter intervals
2. The distance from which mountains of a particular elevation can potentially be seen from sea-level (taking into account light refraction) at 500-meter intervals
3. Altitudes of various reference points across the globe (primarily mountain summits on various continents)
4. Electrical phenomena in the atmosphere (descriptive text)
5. Cultivation of the soil at tropical latitudes (descriptive text)
6. Gravitational force at 1,000-meter intervals
7. Blueness of the sky according to the cyanometer scale (arrayed in elevational zones spanning 1,000 meters each)
8. Air humidity according to Saussure's hygrometric scale (arrayed in 1,000-m zones)
9. Barometric pressure at 500-meter intervals
10. Air temperature in degrees Centigrade in 1,000-meter zones
11. Chemical composition of the air (descriptive text)
12. Height of the lower limit of perpetual snows for various mountains at different latitudes (ranging from 0° to 75°N)
13. Characteristic animals in the Andes, arrayed in 1,000-meter zones (descriptive text)
14. Temperature at which water boils (in degrees Centigrade)
15. Geological perspectives (descriptive text)
16. Light intensity

Taken as a whole, the *Tableau physique* is a colossus of information, representing a wide range of spatial scales. The central panel requires only brief consideration for the viewer to get the sense of systematic pattern in vegetation composition and structure with elevation. At the same time, the viewer can scrutinize any part of the central or flanking panels and obtain a rich and detailed perspective on earth features ranging from the floristics of the Andes to the global-scale bedrock geology of mountains to the temperature and electricity of the atmosphere.

The *Tableau physique*'s representation is by no means perfect, either artistically or scientifically. The asymmetry between the two flanking panels sets the ensemble slightly off balance; it is not a perfect triptych. The text-columns are in no particular order and hence do not come together by themselves to form a story or theme. The floristic and vegetational details in the central panel can be overwhelming, and contain some inconsistencies (e.g., positions of some labels convey precise altitudinal and longitudinal information, while others do not). But overall the *Tableau physique* succeeds in its goals of conveying detailed information in an aesthetically pleasing format, embedding rich details in a holistic framework.[39]

VI

". . . a university, a whole French academy, traveled in his shoes."[40]

The "Physical Tableau" portion of the *Essay*, comprising more than 26,000 words (nearly 80 percent of the total *Essay* text), is a detailed exposition of the Chimborazo figure. Humboldt may well have written the world's longest figure caption. He starts with an extended introduction, including a full description of the central panel of the *Tableau physique*. This is followed by sixteen sections, each corresponding to a column in the panels that flank the central figure. Like the *Tableau physique* itself, the "Physical Tableau" text is rich in detail. It shares the usual features of Humboldt's writing: numerous digressions and cross-references, prose ranging from smooth and lucid to clunky and obscure, and lyrical passages alternating with dry technical descriptions.

The first few pages provide an overview of the *Tableau physique*, its construction, and its aims. Humboldt notes the changes in the vegetation, the fauna, the bedrock, and the air in the ascent from sea level to a high mountain summit, and he observes that the most dramatic changes and broadest elevational variation can be found at the equator. He describes the *Tableau*

39. The *Tableau physique* was developed during the charter era of statistical graphics and abstract visualization in the sciences. Natural history of the early nineteenth century had a long tradition of illustration, but it consisted entirely of detailed, literal representations of plants and animals for descriptive and taxonomic purposes. Scientific works of the eighteenth century often included literalistic depictions of laboratory apparatus or schematics of machinery and devices, which were intended to help the reader visualize or reproduce the measurements or experiments. But formal depiction of abstract concepts or visualization of ensembles of measurements was primitive until the 1801 publication of William Playfair's *The Commercial and Political Atlas* and *The Statistical Breviary*. Graphical representations in Humboldt's later works (especially the *Political Essay on the Kingdom of New Spain*) were strongly influenced by Playfair. Humboldt's 1817 *De Distributione Geographica Plantarum* includes a more compact and elegant representation of vegetational patterns along elevational and latitudinal gradients.

40. Ralph Waldo Emerson, remarks on the centennial of Humboldt's birth in 1869 (see Emerson 1884).

physique as encompassing "almost the entirety of the research I carried out during my expedition in the tropics." As such, it represents a synthesis of Humboldt's "general physics," which builds from "individual studies" to connect "all the phenomena and productions on the surface of the earth." These phenomena include the diversity of life forms and their adaptations to the environment, the contrast between alpine snowfields and volcanic fire at the same high elevations, the marine fossils found on mountain summits, and the properties of the atmosphere observed in Gay-Lussac's balloon ascents above Paris. Such things must surely "seize our imagination" and push us to understand them and their interrelationships. Humboldt admits that he cannot yet connect all the phenomena he describes, nor even characterize them in all the detail necessary. The reader can sense tension between Humboldt's compulsion for synthesis and his recognition that synthesis must rest on a broad foundation of detailed observations. In spite of Humboldt's hundreds of thousands of individual, location-specific observations and measurements, the empirical and conceptual bases required to link all these phenomena remained inadequate.

Humboldt identifies a second tension more explicitly, namely the trade-off between aesthetics and precision in developing the *Tableau physique*. In places he seems defensive or apologetic in describing how the figure was constructed. This is understandable: the profile was an entirely new sort of graphic, combining elements of traditional cartography, botanical illustration, landscape profiles (like La Condamine's), and statistical graphics (still in its infancy). Humboldt was groping not only toward intellectual synthesis but toward conveying his synthesis in some effective and pleasing visual representation.

Humboldt takes great pains to explain the empirical foundation for the figure—the portion of the globe it was intended to represent, the rationale for inclusion of various features, the precise geographic features the profile was designed to depict, the nature of the various observations included in the figure. He discusses the diverse estimates of Chimborazo's summit elevation[41] and the reason for portraying Cotopaxi instead of some volcano closer to Chimborazo.[42] He grapples with the issue of vertical exaggeration, and in-

41. The true elevation is 6,268 meters. La Condamine's estimate was the closest of those Humboldt discusses, though this may represent luck more than superior accuracy or precision.

42. Humboldt's bizarre reference to fish-spewing volcanoes on page 85 (and in the *Tableau physique*) was repeated in his essay on volcanoes in *Views of Nature* and discussed

troduces a thought experiment to show how ridiculously small Chimborazo would become if portrayed to the same vertical and horizontal scales.

The second half of the introductory text is devoted to plant geography, first describing the empirical support for the figure and explaining its rationale and deficiencies of detail. He then commences an elevational tour of Andean vegetation, starting with subterranean fungi and submarine algae, and then moving up the slopes from sea level to the Chimborazo summit. Emphasis is on physiognomy and floristic composition of vegetation, with occasional discussion of elevational limits of taxa or growth-forms and their environmental tolerances. He takes the reader on several digressions, particularly on plants of particular economic[43] or biogeographic[44] interest.

Once Humboldt reaches the snowfields of Chimborazo, the tour is complete. He then turns to Europe, noting how useful a comparable *Tableau physique* for temperate climates would be. In spite of the extensive botanical work done in Europe, the published floras lacked the detail and spatial precision of the records of Humboldt and Bonpland.[45] Though Humboldt stops short of developing a second *Tableau* for Europe, he has no inhibition about providing a verbal sketch, using Mont-Blanc as the central component. Having ascended the slopes of the Andes to snowline, he descends from the snows

---

in detail in a note ("Mémoire sur une nouvelle espèce de Pimelode, jetée par les volcans du royaume de Quito") published as part of his 1811 zoological monograph. In the latter work, he admits that although he never witnessed this phenomenon himself, it was "such a common phenomenon and so generally known among the local people, that there cannot be the least doubt of its authenticity" (my translation). Humboldt respected the accounts and opinions of local European and Creole inhabitants as well as indigenous people he encountered in his travels. Nevertheless, his credulity in this affair is surprising. Undoubtedly, volcanic muds rich in hydrogen sulfide smelled something like dead fish, but there is no evidence that fish occupied subterranean lakes under the volcanoes as Humboldt concluded (see Romero and Paulson 2001).

43. These include the various gum- and latex-producing plants (e.g., rubber), as well as important medicinal plants (e.g., *Cinchona* or Peruvian-bark). Humboldt took a particular interest in plants of practical use (see Humboldt's 1820 work on *Cinchona*, as well as extended passages in the *Personal Narrative*).

44. Biogeographic discussions include species of South America that do not occur beyond the Panama isthmus (e.g., *Cinchona*, *Cheirostemon*), and some that occur in Mexico and North America (e.g., *Quercus*). Humboldt discusses these and other biogeographic patterns at greater length in digressions in his *Personal Narrative*.

45. This may explain Humboldt's exhortations in the opening paragraphs of the "Essay."

of the Alps to the sea, with the usual excursions to point out unusual features and phenomena.

Although a few comparisons are drawn between the European and Andean patterns, Humboldt does not compare the continents systematically.[46] Perhaps he perceived a need for additional information and detail, from Europe and elsewhere. A sense of his global vision comes near the end of the introduction, when Humboldt makes an appeal for a global array of comparable tableaux—elevational patterns of vegetation and other phenomena from different latitudes, different hemispheres, different continents. Here he returns, in gentler form, to the exhortation of the "Essay": botanists must strive not only to collect and describe plants but to put them in geographic and environmental context in a systematic way—to treat them ecologically. Humboldt goes so far as to mention names of individual botanists who could provide such tableaux for different regions of the globe.

The remainder of the "Physical Tableau" text—some 60 percent of it—is a column-by-column explanation and description of the sixteen panels of the *Tableau physique*. They follow a more logical order than the *Tableau physique* itself. The first ten focus on physical and chemical properties of the earth and its atmosphere. Most share a common structure, in which Humboldt describes elevational patterns, then lateral (especially latitudinal) patterns, intermittently digressing on various topics of interest. These latter are often on vegetation and flora but also include geological, zoological, and cultural patterns. Humboldt points out unique features of plant morphology and physiology in response to variations in temperature, air pressure, humidity, and light intensity. Other departures from the central descriptions center on problems of instrumentation and measurement in field settings.

In a long section devoted to "Geological Considerations," Humboldt observes that, in contrast to vegetation, climate, or other features of the earth, bedrock composition and structure show no consistent or universal pattern with elevation. "Primary" (igneous and metamorphic rocks) and "secondary" (sedimentary rocks) can occur anywhere, from sea level to the highest peaks. Individual mountain ranges and continents may show predictable

46. Goethe filled the breach in part; after reading a draft of Humboldt's *Essay*, Goethe drafted a profile comparing the "ancient continent" with the "new continent." This profile, reproduced at the end of this essay, shows Mont-Blanc on the left (with Saussure on the summit) and Chimborazo on the right (with Humboldt at the appropriate elevation). The margins are scaled altitudinally, and include various features of interest (mountain summits, cities, etc.).

patterns, but there seem to be no general laws, at least as far as elevation is concerned. Humboldt describes the geology of the Andes in detail and proposes that the Americas are flanked by mountain ranges extending from southern Chile to Alaska.[47] Humboldt briefly notes similarities in stratigraphy among different continents.[48]

After brief sections on global snowline elevations and the distance mountains can be seen from the sea, Humboldt summarizes zoological patterns along the Andes slopes. He makes no global comparisons, in contrast to the botanical discussion, but he again notes the potential for a series of zoological tableaux from various points on the globe to parallel the botanical tableaux. He then turns to agriculture and human societies. Again the focus is strictly on the Andes. The influence of Johann and George Forster is obvious in Humboldt's treatment of environmental influences on culture. Of particular interest is Humboldt's influential argument that "the civilization of peoples is almost always in inverse relation to the fertility of the soil they occupy." He cites the Inca culture specifically, comparing it with the hunter-gatherer cultures in the Amazon and Orinoco lowlands to the east. He had no way of knowing, of course, that sophisticated agricultural societies occupied the Amazon lowlands until the early sixteenth century.[49]

The "Physical Tableau" text concludes with an extensive table of elevation

47. Humboldt is most vague about the section of North America that is now the western United States. At the time, the geography of this region was almost completely unknown in Europe or the United States. The first clear indication that the continent was spanned by a broad and complex series of mountain ranges came from the experience of the Lewis and Clark Expedition. They did not return to St. Louis until September 1806, and reports did not reach Europe until 1807, after the *Essay* had gone to press.

48. Humboldt summarized his geological thinking in *A Geognostical Essay on the Superposition of Rocks, in Both Hemispheres*, published in 1823. He also speculated extensively on intercontinental geology in the *Personal Narrative*.

49. The extent to which Amazonia was cleared and cultivated in pre-Columbian times remains controversial. However, there is consensus, based on archeological and paleoecological studies of the past few decades, that forest clearance and agriculture were widespread in the region. Several scholars have argued that the entire Amazon landscape was "cultural"—i.e., intensively managed—in pre-Columbian times. Their arguments are ably summarized in Charles Mann's book, *1491*. Some ecologists and paleoecologists, however, have uncovered evidence that cultural impacts were heterogeneous, ranging from intensive along the major river arteries and other regions to minimal in many interfluves. Mark Bush and Miles Silman present the case for cultural-

measurements worldwide, including cities, mountain summits, and mountain passes. A brief two-part addendum discusses some corrections to measurement conversions and some recent botanical and zoological information from the United States.

Viewing the *Essay on the Geography of Plants* as a whole, the *Tableau physique* is far and away the most successful and effective of the three components. Not only does it convey complex concepts and information in the universal language of graphics, it also comes closest to achieving Humboldt's holistic goal of synthesis, letting general patterns emerge from fine details. The text portions of the work—the "Essay" and the "Physical Tableau"—fall short in achieving synthesis. The "Essay" lays out the pieces required to assemble a general geography of plants (climate, human history, etc.), but it never fits them together. The "Physical Tableau" text is highly detailed throughout, but there is little discussion of pattern emerging from these details, either within any of his seventeen variables or among them.[50]

The shortcomings of the *Essay* are understandable. Humboldt set out to synthesize a huge amount of information in less than a year's time after his return from the Americas. His seemingly paradoxical observation in the preface that "if I could have worked on this book for a longer time, it would have been even shorter" (61) makes sense in this light. The task of simply compiling and describing all of the information he had assimilated was gargantuan enough. Truly synthesizing it—reducing it to a set of general patterns and principles—would require far more time than he had. Only in his pictorial representation did he truly succeed in consolidating an abundance of particulars into a few generalities.

VII

Chimborazo is the centerpiece of the *Tableau physique*, and the *Tableau physique* is the centerpiece of the *Essay*. The *Essay* was the first and most synthetic work Humboldt prepared from his travels. And it was a synthesis that was well underway in Humboldt's mind before he departed from Europe. Yet his

---

impact heterogeneity in recent articles in the *Royal Society Proceedings* and in *Frontiers in Ecology and the Environment*.

50. I include plants as the seventeenth in addition to the sixteen panels.

ascent of Chimborazo, in fact his entire sixteen-month exploration of the Andes, was unplanned. Before embarking for Cartagena in his attempt to rendezvous with Baudin's expedition, Humboldt was far more interested in exploring the interior of North America, the highlands of Mexico, and the islands of the South Pacific. And after his voyage, he had the choice of any number of mountains on which to center his synthesis: Vesuvius or Mont-Blanc in Europe, Pico de Teide in Tenerife, the Silla near Caracas, Popocatépetl in Mexico. How did Chimborazo come to be the focal point of Humboldt's synthesis?

We can identify plenty of objective reasons for Humboldt's choice. Chimborazo is less than two degrees off the equator, and Humboldt recognized that the broadest elevational gradient on earth—in climatic terms—should be in the highest equatorial mountains. Permanent ice and snow occurred at Chimborazo's summit, and the warm end of the gradient was not truncated at sea level, as it would be at higher latitudes. Humboldt was searching for archetypes, and a global archetype for climate and elevation required both end members—hot and cold, high and low. Chimborazo, reputedly the highest summit in the world and located conveniently close to the equator, provided the full spectrum of elevation and temperature.

Furthermore, selection of a tropical mountain as the global archetype made it easy to draw the analogy between latitudinal and elevational gradients in climate and vegetation. The parallel changes as one ascended the slopes of Chimborazo or traveled from the equator to the poles provided a strong argument for the primacy of climate in controlling vegetation composition and structure. Humboldt drew this connection in the "Physical Tableau" portion of the Essay, where he turned from the Andean vegetation tour to his account of Mont-Blanc and the Pyrenees. He made the connection more explicit in De Distributione Geographica Plantarum (1817), where he included a pictorial representation of three vegetational profiles: equatorial Chimborazo, mid-latitude Mont-Blanc, and Sulitelma in arctic Lapland.[51] It was this graphic that finally connected the dots between altitude and latitude, coming closer to his Essay vision of a global array of tableaux physiques.

Humboldt was pondering climate, elevation, latitude, and vegetation well before his Andean travels, in fact well before he set out for his travels outside Europe. Karl Ludwig Willdenow had already set him to thinking about

51. A black-and-white reproduction of the 1817 diagram follows this essay, along with another Chimborazo profile published as part of the Personal Narrative.

latitudinal and altitudinal analogies, and George Forster had drawn his attention to the role of climate in controlling vegetation and culture. But that was all secondary information, book-learning. Humboldt had traversed the Alps with von Buch and the Pyrenees with Bonpland, witnessing the floristic changes and feeling the climatic changes firsthand. He was already planning some sort of monograph on plant geography. But his global vision of climate and vegetation was still incoherent—otherwise he would have identified the highest equatorial mountains as a destination when he left Europe.

Humboldt's ascent of the Pico de Teide on Tenerife undoubtedly reinforced his previous thinking based on his experience in the mountains of Europe. His ascent of the Silla of Caracas would have done the same. But the Silla was simply not high enough to reveal the full gradient, particularly to draw analogies between the alpine zones of the tropics, the Alps, and the Arctic.

Humboldt's thinking crystallized during his travels in the Andes. He was surely stunned to see the vegetation patterns and snowline in the mountains firsthand, particularly after two years in the tropical lowlands. He also had the advice and assistance of Mutis and Caldas, who knew the territory and its plants. They played important roles in Humboldt's thinking, and the detailed botanical chart in the *Tableau physique* relied in part on their observations. But Humboldt had the foundational ideas before he arrived in the Andes. All that was missing was direct experience in tropical mountains to connect and conceptualize them.

Humboldt wrote the first draft of the *Essay* and prepared the sketches that evolved into the *Tableau physique* while in the vicinity of Chimborazo. He also wrote a draft of his essay on "Ideas for a Physiognomy of Plants," later published in *Views of Nature*, while in the Andes. Surely something powerful motivated him during this creative period. Although some scholars have suggested that Humboldt was motivated by fear of being "scooped" by the locals, particularly Caldas, this seems unlikely. Humboldt's ideas were far better developed, and his botany was more sophisticated, than Caldas's.[52] He was undoubtedly stimulated by his discussions with Caldas and with Mutis. But much of the inspiration must have come from his having just experi-

52. Humboldt also sent an early draft of the *Essay* and the *Tableau physique* sketch to Mutis for comment, and Caldas reprinted the draft in the scientific journal he edited (Appel 1994, 58, 81–82). These hardly seem like the actions of people competing for priority.

enced the gradients firsthand. Humboldt caught a glimpse of the cosmos in the high Andes. Synthesis became inevitable.

VIII

"The enormous popularity of [Cosmos] among the educated of the entire Western world, marked it out . . . as the standard of science literacy."[53]

Humboldt's best-known and most widely read work was his five-volume *Cosmos: A Sketch of a Physical Description of the Universe*, published between 1842 and 1862. The work was published repeatedly in the nineteenth century in German, French, English, and several other languages. The first two volumes remain in print today in English translation. Humboldt delivered a series of lectures on the subject in Paris and Berlin between 1825 and 1828 but later confessed that he began outlining the work in 1819.[54] Late in his life, Humboldt asserted that the work derived not from the Paris/Berlin lectures but from his travels in South America. And he was discussing plans for a *physique du monde* as early as 1793.

Cosmos, an ambitious work that Humboldt labored on until his death in 1859, was conceived as a synthesis of the known world, ranging from the interior of the earth to the earth's surface features, biota, and atmosphere, to the extraterrestrial phenomena of the solar system and stars. It was not intended to encompass the universal principles of physics or chemistry—such a scope was beyond even Humboldt—but rather to apply them to explain the observable properties of the world we inhabit. Thus, the work focused not on physics but on geophysics, climate, and astronomy, and not on chemistry but on mineralogy, geochemistry, and cosmochemistry. Likewise, Humboldt did not treat the emerging fields of biology and physiology in *Cosmos*, but applied them to biogeography and ecology.

Originally conceived as a two-volume work, *Cosmos*, like Humboldt's *Personal Narrative*, turned into a prolonged and ultimately unfinished series. Volume 1 is a 360-page introduction and synopsis, while volume 2 comprises

53. Nicolaas A. Rupke, Introduction to the 1997 edition of Alexander von Humboldt, *Cosmos*, 2 vols. (Baltimore: Johns Hopkins University Press, 1997), 1:xxxii.

54. The history, development, and reception of *Cosmos* are summarized in introductory essays by Nicolaas Rupke and Michael Dettelbach to the 1997 editions of the first and second volumes and in Rupke 2005.

a comprehensive summary of the history of human interest in and appreciation for the natural world. Astronomy is treated in detail in volumes 3 and 4, which respectively summarize outer space and the solar system. The unfinished volume 5 is centered on geophysics of the earth. Humboldt never completed the anticipated volumes on the earth and its inhabitants.

Humboldt's *Essay* was the critical intermediate between Humboldt's early conception of a comprehensive *physique générale du monde* (general physics of the world) and its partial realization in *Cosmos*. In essence, the *Essay* is *Cosmos* without the extraterrestrial elements of volumes 3 and 4 and the philosophical and historical background of volume 2. Because Humboldt never completed the terrestrial portions of *Cosmos*, the only glimpses we have of his comprehensive vision must come from the *Essay* itself, and from the summary in volume 1 of *Cosmos*, of which some thirty pages are devoted to life on earth.

The narrative thread of *Cosmos*' "General Review of Natural Phenomena" runs from the celestial to the cultural. Humboldt reviews the observable phenomena of the day and night sky and then turns his attention to the earth, treating first its interior ("the inner life . . . of the earth"), and then its surface and atmosphere. He returns to the various physical phenomena treated in the *Essay* but places them in a broader, more integrated framework. The fruits of thirty-five years of progress, both in accumulation of scientific knowledge and in Humboldt's own thinking, are apparent in *Cosmos*. Humboldt finds himself in better position to make sense of spatially dispersed sets of simultaneous observations (of magnetism, of earthquakes, of ocean and air temperatures, of wind and rain, all accumulated since 1807) and to draw causal connections among diverse phenomena. He outlines thought experiments, wherein the earth is imagined as a series of alternative planets—one with a universal ocean, another with a uniformly flat land mass, another with the Americas oriented longitudinally rather than latitudinally—in order to explore the effects of ocean and continental configurations on climate. Humboldt's *physique du monde* can be glimpsed as he discusses the influence of ocean currents on the atmosphere, the emission of carbon dioxide by volcanoes and living organisms and its assimilation by plants, and the interrelations among sunlight absorption, heat emission, atmospheric pressure, humidity, and the surface of the earth.

Humboldt's treatment of plant geography in *Cosmos* is brief, lacking the extended discussions of earlier sections. Although this may stem from his having provided extended treatment in his 1817 *De Distributione Geographica Plantarum*, it may also reflect the sheer magnitude and diversity of the topic.

He reiterates his assertions from the Essay concerning the primacy of vegetation in human perceptions of the landscape, and notes the importance of climate in determining distributions of species and growth forms across the earth's surface. Perhaps his most important generalization, one hinted at in the Essay, is that plant form and vegetation physiognomy vary more systematically with geography and environment than do taxonomic groups such as genera and families. This was, of course, a key puzzle solved by Darwin: natural selection calls forth growth forms best suited to local environments; geographic distributions of taxonomic (phylogenetic) groups have strong historical components.

Volume 1 of Cosmos concludes with a brief discussion of the distribution of human races and languages, a topic not treated in the Essay. Humboldt argues that humans comprise one species, that human cultures are shaped by their environments, and that all humans are entitled to freedom and dignity. He ends with an eloquent summary of the scope of Cosmos, observing that phenomena from the celestial to the terrestrial can be arranged "according to partially known laws," and he states that the "higher spheres of the organic world," including the human mind, are governed by "other laws of a more mysterious nature." Humboldt anticipates syntheses beyond what he has been able to provide, but he cannot envision what form they might take. It remained for Darwin to provide the first such synthesis for biology and behavior.

IX

"He was also, in a sense, a victim of his own success: his presence everywhere meant that he was nowhere in particular, leaving no field or school to bear his name. . . ."[55]

Humboldt's Essay solved few outstanding problems, either in plant geography or more broadly in the science of the earth. Its major contribution lay in delineation of a broad, integrative vision of science and the kinds of information required to fulfill that vision. Humboldt set the course of much of nineteenth-century science by identifying the overarching questions of the earth's surface, interior, and atmosphere, demonstrating that these questions were interlinked, and by developing tools needed to address them. These tools included: (1) precise and accurate measurements of physical and chemical properties and geographic location, (2) dispersed spatial arrays of

---

55. Laura Dassow Walls 2001, 128–29.

simultaneous measurements, ranging from floristic inventory to temperature and geomagnetism, (3) graphical displays of quantitative and qualitative information, including iso-maps and geographic profiles, (4) search for
general theoretical and conceptual frameworks to explain the particulars
within a field of knowledge (e.g., average air temperature), and (5) pursuit
of formal conceptual relationships among diverse phenomena (e.g., how
latitude, temperature, physiography, humidity, wind direction, snowfall, and
proximity to the ocean influence the lower limit of permanent snow across
the globe). Humboldt's vision set in motion many of the great scientific advances of the early to mid-nineteenth century.[56]

Systematic exploration of the world's oceans and continents through the
nineteenth century applied the tools and philosophy advocated by Humboldt
and led to rapid documentation of floras, faunas, landforms, ocean currents,
climate patterns, and native cultures across much of the globe. These explorations were often formally associated with national or imperial activities,
most had explicit economic or military motivations, and most were designed
along Humboldtian lines, with careful attention to precise measurements,
geographic context, and extensive collections.[57] An important feature of
these explorations was the integration of scientific with practical objec-
tives. Participants, as voyagers or curators, included such scientific figures
as Charles Darwin, Thomas Henry Huxley, John Torrey, Asa Gray, Edward
Forbes, Matthew Fontaine Maury, David Dale Owen, and Peter Kropotkin.
Other expeditions were private affairs, bankrolled by wealthy individuals
with interests in natural history (Louis Agassiz's explorations of Lake Su-

56. I am indebted to Susan Faye Cannon's 1978 *Science in Culture*, particularly chapter 3,
for many of these insights. Michael Dettelbach's writings on Humboldtian science have
also been very helpful. I also acknowledge Malcolm Nicolson's comprehensive works
on Humboldt's links to modern ecology.

57. William Goetzmann has written extensively about these expeditions, particularly
in the United States, and noted Humboldt's influence on organization and execution.
As one example among many, Congress mandated Secretary of War Jefferson Davis to
organize the Pacific Railroad Survey of 1853–55. Davis dispatched military engineers
on four longitudinal surveys from the Mississippi to the Pacific. A fifth survey ran latitudinally, from San Diego to Seattle. Each survey party was accompanied by soldiers,
cartographers, artists, botanists, and other naturalists, and each collected extensive
plant, animal, mineral, and fossil specimens. The project yielded thirteen profusely
illustrated volumes with extensive astronomical data for cartography, climate measurements, illustrated monographs on flora and fauna, descriptions of native cultures, and
topographic profiles and maps showing geology, plant distributions, and forest cover.

perior and the Amazon,[58] and the Harriman expedition to Alaska number among these).[59] A few others were entrepreneurial undertakings, aimed at providing specimens for seemingly insatiable domestic markets in exotic plants, butterflies, bird eggs, minerals, and other curiosities. Alfred Russel Wallace and Henry Bates launched their scientific careers this way, explicitly inspired by Humboldt's travels.

Although many of these early to mid-nineteenth century surveys led ultimately to establishment of more or less permanent government agencies in the late nineteenth century (e.g., the U.S. Geological Survey, various state geological and natural-history surveys), the grand, comprehensive enterprises became less frequent and more local to regional in scope towards the end of the century. This was partly a matter of running out of unexplored territory, but it also stemmed from the inevitable atomization of broad-based natural science into specialized disciplines. By the late nineteenth century, Humboldt's intellectual descendants were dispersed among subdisciplines within geology, botany, zoology, climatology, oceanography, taxonomy, and other fields.

By the turn of the twentieth century, Humboldt's ideas on precise measurement, spatial arrays of data, graphical and cartographic portrayal of information, and pursuit of general conceptual frameworks from particular details were deeply embedded in all of the disciplines concerned with the study of the earth, including its surface and atmosphere. Synoptic networks and cartographic display were important features of twentieth century geology, climatology, and ecology. Pioneering ecologists, including C. Hart Merriam, Charles C. Adams, Henry Chandler Cowles, Frederic Clements, Henry Allen Gleason, Forrest Shreve, and Victor Shelford, recognized the overrid-

58. Agassiz's 1848 canoe excursion along the north shore of Lake Superior was organized as a field course, funded by subscription of wealthy students from Harvard and other Boston-area schools. Agassiz and party described new features of the geology, botany, and zoology of the region. The 1865 expedition to Brazil, in which Agassiz was accompanied by his wife, several other scientists, and a young William James, was a comfortable affair, underwritten by Boston financier Nathaniel Thayer.

59. The 1899 Harriman Expedition was sponsored by railroad magnate Edward Harriman, who chartered a steamship to carry G. K. Gilbert, C. Hart Merriam, George Bird Grinnell, John Muir, John Burroughs, William Trelease, and sixteen other scientists as well as artists (including Frederick Dellenbaugh and Louis Agassiz Fuertes), photographers (including Edward Curtis) and members of Harriman's family along the coast of Alaska. Looking Far North, by William Goetzmann and Kay Sloan, is an excellent account of this all-star expedition.

ing role of climate in determining plant and animal distributions as well as the physiognomy of vegetation. Curiously, most of these ecologists and their contemporaries avoided abstract visualizations like Humboldt's *Tableau physique*.[60] Ecogeographic abstractions returned in the 1950s and soon became widespread in the ecological literature.[61] These latter visualizations were not without controversy, arising from the same tension within ecology that Humboldt felt within himself: literal accuracy and precision on the one hand, and clarity of representation and communication on the other. Some twentieth-century ecologists openly criticized such visualizations, even more concrete maps, because they inevitably smoothed over fine-scale spatial details and nuances.[62]

60. Merriam's 1890 elevational profiles of the San Francisco Peaks region are a notable exception. However, Merriam was of an earlier generation than the other ecologists mentioned here. Until the 1950s, illustrations in ecological literature consisted primarily of statistical and time-series plots of data, maps (often showing vegetational patterns at scales ranging from 10^0–10^{10} m^2), and photographs or line drawings. The latter were typically direct representations of morphology or anatomy of organisms, or scaled-down, maplike depictions of field phenomena (e.g., soil profiles). Graphical abstractions were largely confined to box-and-arrow schema showing successional pathways and time-series of population dynamics from mathematical models. Topographic profiles in ecology, with rare exceptions (e.g., Rasmussen 1941), were restricted to highly localized gradients (e.g., transects from a stream valley to an adjacent ridge, or from shallow water to well-drained uplands—all within the space of 10^2–10^3 m). The apparent dearth of topographic and other Humboldtian graphical abstractions in ecological literature of the early twentieth century deserves further study.

61. Topographic and other Humboldtian visualizations appeared in ecology during the early 1950s with papers by W. Dwight Billings, Robert H. Whittaker, and John T. Curtis and students. Initial representations were of vegetation at broad geographic scales, similar to Humboldt's profiles and tableaux. However, they were soon adapted to other spatial contexts. For example, Robert MacArthur's idealized spruce-tree profiles showing feeding areas of various warbler species, and Joseph Connell's intertidal profile of barnacle ecology might be considered Humboldtian abstractions.

62. The history of ecology is summarized in works by Donald Worster, Robert McIntosh, Sharon Kingsland, Peter Bowler, and Joel Hagen. These are all important and insightful contributions, but they leave plenty of room for deeper analysis of the role of philosophical tensions, prevailing norms of visual representation, and interactions with other disciplines in the development of ecological thinking. These issues all have roots in Humboldt's time and thinking. For example, Humboldt's polarity between physiognomic and taxonomic/compositional approaches to vegetation remains a tension in plant ecology today. As a biological science, ecology has centripetal forces pulling it towards a focus on biological processes, internally driven dynamics, and phy-

Humboldt's descendant sciences were eclipsed through much of the twentieth century by rapid advances in physics and physiology, both initiated by rigorous laboratory experimentation. The inferential power of controlled experimentation led to establishment of physiology and its derivatives (e.g., cell biology, biochemistry, molecular biology) as the standard-setting disciplines in the life sciences, with parallel emphasis on physics and chemistry in the physical sciences. Those standards, at least as perceived by many practicing scientists, were not only methodological (experimental inference) but epistemological (emphasis on reductionism and determinism, with a search for universal laws). The earth, atmospheric, and ecological sciences continued to progress but were often perceived as secondary, descriptive sciences. Tensions and even schisms developed within these sciences in the late twentieth century as they struggled over whether to model themselves after the "superior" disciplines. For example, ecological controversies in the 1970s and 1980s centered on whether universal ecological laws might exist, and on the relative merits of experimental versus observational approaches to scientific inference.[63] These tensions in the sciences of the earth are ironic in view of Humboldt's simultaneous advocacy and practice of both experimental science and synoptic, descriptive science.

The past few decades have witnessed a convergence of several separate disciplines toward a new incarnation of Humboldt's *physique du monde*. An early phase of this instauration took place in the earth sciences in the 1960s

logenetic frameworks. At the same time, as an environmental science, ecology is subject to centrifugal forces that lead to alternative focus on physicochemical processes, external drivers of ecological dynamics (e.g., climate variability), and organizational frameworks based on form and ecosystem function rather than phylogeny. These and other tensions have influenced the history of ecology since Humboldt's time and are worthy of closer examination.

63. These issues and controversies are much more diverse and complex than I can portray in a few sentences here. Methodological and epistemological controversies developed in ecology, evolutionary biology, and many of the earth and atmospheric sciences in the late twentieth century. These controversies tended to align themselves along similar axes—experimental to observational, synoptic to localized, holistic to reductionistic, universal to particular, historical to ahistorical, deterministic to stochastic, internal to external loci of explanation. Unfortunately, these polarities were often treated as dichotomies by the various protagonists. Most of the relevant subdisciplines seem to have gotten beyond these issues in the past decade or two. The tensions remain, but most often in a spirit of *détente* or creative, dialectical engagement rather than hot or cold warfare. These tensions and their resolution deserve attention from historians and philosophers of science.

and 1970s, with the development of plate tectonic theory, providing unified explanations of disparate geophysical, stratigraphic, paleontological, and biogeographic observations. The plate tectonics revolution was made possible by extensive spatial arrays of observations—geomagnetic, paleomagnetic, seismological, stratigraphic—developed along the lines envisioned by Humboldt.

A second, even more sweeping unification of the sciences of the earth is underway, driven by a unique intersection of scientific advances and societal needs. This unification brings the geophysics of plate tectonics together with processes of the solar system, oceans, atmosphere, land-surface, ecosystems, and societies. The scientific activity has been driven by societal concerns about global climate change and by the dawning recognition in the 1980s and 1990s of the diverse and complex controls of atmospheric circulation. Climate is influenced by solar radiation, continental configurations, and ocean circulation, as Humboldt recognized. But we know now that continents move via plate tectonics, that ocean circulation patterns undergo change,[64] and that solar radiation has varied owing to the gravitational effects of neighbor planets,[65] sunspot variation, and changes in the amount of dust and aerosols in the atmosphere. We now recognize the effects of atmospheric chemistry (including greenhouse gases such as carbon dioxide and methane), vegetation cover, and human land-use on climate at regional to global scales. We have learned that greenhouse gas concentrations are influenced at various time scales by tectonic activity, rock weathering, sequestration in soils and sediments, combustion of fossil fuels, absorption by

64. Such shifts occur across a wide range of timescales, from interannual (e.g., El Niño/Southern-Oscillation [ENSO] variation) to multimillennial (e.g., effects of insolation variation, ice-sheet dynamics, and sea level) to changes wrought by moving continents over millions of years.

65. Orbital (a.k.a. Milankovitch) variation, by which the Newtonian celestial mechanics of planetary orbits influence each other, has been recognized as a potential force on earth's climate since the mid-nineteenth century (see Imbrie and Imbrie 1979). Interactions of earth's orbit with those of other planets (particularly Jupiter and Saturn) lead to systematic, cyclic variations in several parameters of earth's orbit, which cause changes in the seasonal and latitudinal distribution of solar radiation. This variation is in part responsible for the glacial/interglacial variation of the past million-plus years, as well as higher frequency (millennial to multimillennial) climate variation. William Ruddiman's newly revised text, *Earth's Climate: Past and Future*, reviews Milankovitch variation as well as other factors influencing climate, ranging from tectonic uplift to prehistoric agriculture.

ocean waters, fixation by vegetation, release by wildfires, and industrial synthesis of novel compounds (e.g., chlorofluorocarbons). We have discovered
important feedbacks, negative and positive, among the various components
of the system—including some anticipated by Humboldt, who speculated
that effects of vegetation physiognomy on wind velocity, light reflection, and
evaporation could affect climate at a regional scale.[66]

These are just a few of the elements required for a full understanding of
how the earth's climate system is governed, how it has varied in the recent
and geologic past, and how it is being affected and will continue to be affected by human activities. Understanding all the relevant processes and
interactions, across a vast range of temporal and spatial scales, requires synoptic measurement arrays from local to global, records of dynamics and responses from real-time to near-time to deep-time, articulation of processes
from a broad array of phenomena, and cross-disciplinary communication
and collaboration unprecedented in the recent history of science. This is
a mammoth enterprise—not unlike Humboldt's enterprise of a *physique
générale du monde.*

The new climate-oriented earth-science enterprise got underway in the
1980s, developed momentum in the 1990s, and has continued accelerating
as fundamental new phenomena and connections are discovered and as new
disciplines are drawn in. The new science is young enough to lack a universally recognized name, though "integrated earth sciences" and "earth system
science" have been touted. Perhaps "geophysics" would serve as well as any.
The scientific organization that best encompasses the ongoing integration
of the earth sciences, including the ecological and most recently the social
sciences, is the American Geophysical Union. And Humboldt's *physique du
monde* translates as "physics of the earth." All that is needed is to expand the
conception of "physics" back to its Enlightenment definition: the study of
the natural or material world—a.k.a. "science."

The unification of the sciences of the earth comes not a moment too soon.
Human activities—land clearance, vegetation conversion, transcontinental
movement of species, harvesting of resources from lands and waters, industrial production of biologically active nitrogenous compounds and most
especially greenhouse-gas emissions via combustion of fossil fuels—are

66. Humboldt discussed effects of vegetation cover on climate in his essay "Ideas
for a Physiognomy of Plants" in *Views of Nature* (see especially 216–17 and footnote 9
in the Bohn translation) and in vols. 4:142–45, 337 and 6:7–8, 62–66 of the *Personal
Narrative.*

placing the ecosystems and human societies of the world in grave danger of disruption, even extinction, in the coming decades. Nothing less than an integrated science of Humboldtian scope is required to determine the damage already done, the risks we still face, and the options for prevention and mitigation.

X

Humboldt's last portrait, painted by Julius Schrader a few months before his death, shows him as a pensive, white-haired scholar, sitting slightly bent over a notebook, with Chimborazo prominent in the background. The snowline is conspicuous just above Humboldt's right hand. The overall contour of the mountain, and the level of snowline, is modeled after an 1810 painting by Friedrich Georg Weitsch. Chimborazo dominates Weitsch's painting, with Humboldt and Bonpland as small figures in the foreground, along with native people, plants, instruments, and specimens. The Schrader painting conveys Humboldt as the elder statesman of science. He is no longer the youthful field scientist, sextant in hand. And he has grown in proportion; he dominates the painting. Chimborazo, though reduced, remains a timeless, unchanging backdrop to Humboldt's life.[67]

Humboldt's portrayal of Chimborazo in the *Tableau physique* is equally timeless. The smoke venting from Cotopaxi's cone hints at dynamic forces underlying the scene. Hidden processes occur, determining the specific elevation of snowline, the contour of the mountain, and the array of species growing on its flanks. Humboldt perceived the surface patterns as manifestations of a dynamic equilibrium among various physical and biological forces, an equilibrium that might have persisted for many thousands of years and that could persist for thousands more. He knew that the earth had history, and that Chimborazo might not have always been as he saw it, but he had no way of knowing whether the patterns he saw were hundreds, thousands, or millions of years old.

We no longer live in Humboldt's static world, at least as far as earth history is concerned. As of the early twenty-first century, we know that we have inherited a planet that emerged from an ice age only 15,000 years ago, the

67. According to Halina Nelkin (1980), Humboldt chose the background for Schrader's portrait. Black-and -white reproductions of the Weitsch and Schrader portraits follow this essay.

most recent of more than a dozen glacial/interglacial alternations of the past two million years. Our species developed agricultural technologies on several different continents during the past 10,000 years, resulting in widespread land clearance and conversion long before recorded history. Changes in climate since the ice retreat have driven advance and retreat of species ranges, increases and decreases in population sizes of various species, and changes in the overall appearance and function of landscapes across the globe.[68]

Humboldt's tropics are no exception. During the last glacial maximum, glaciers extended down to 3,600 meters elevation on the slopes of Chimborazo, nearly 1000 meters below the modern limit of permanent ice.[69] They retreated upslope as the earth passed out of the last glacial period and have undergone minor advances and retreats until recently.[70] Similarly, patterns of vegetation in the Andes and the lowland tropics have changed dramatically since the last glacial period. Vegetation zonation patterns along the Chimborazo slopes would have been moderately different had Humboldt visited 5,000 years ago and radically different had he been there 20,000 years ago.[71]

Just as they have changed in the past, the Andes and the rest of the world will change in the future—all part of the natural dynamic of our planet. However, a new dynamic is in play. Industrial combustion of fossil fuels, together with forest clearance and fire, are driving a steady increase in carbon dioxide concentrations in the atmosphere—the core dynamic of global

68. Ecological effects of past climate changes have become an important research focus in ecology, since we have few observational records of how ecological systems actually respond to climate changes. The literature in this field is vast. Some of my own recent papers discuss geographic range-shifts and other ecological consequences of climate change (see Jackson 2004, 2006, and Jackson and Overpeck 2000 and references therein). Many species ranges are still adjusting to climate changes of the past few centuries, and many vegetation patterns are just a few hundreds to thousands of years old.

69. The glacial history of the Andes, including Chimborazo, has been recently summarized by Smith et al. (2005). Clapperton and McEwan (1985) provide descriptions of the Pleistocene moraines of Chimborazo.

70. Lonnie Thompson and his colleagues have done extensive studies of Andean glaciers, obtaining ice-cores that reveal regional climate changes and glacial dynamics. The most recent summary of his studies is in a 2006 review, which includes an alarming status report on Andean glaciers he has studied since the 1970s.

71. Vegetation and climate history of the Andes and other tropical regions are summarized most recently in books by Morley (2000) and Bush and Flenley (2007). Hooghiemstra and Van der Hammen provide a synopsis of their long pollen records spanning multiple glacial cycles from the Colombian Andes in a 2004 paper.

climate change.[72] Global climate change is underway, from the Andes to the polar regions. The snows of Chimborazo are disappearing. Glaciers all over the Andes are melting at an alarming rate, with potentially grave consequences for human societies in the region.[73] Ecological changes to date are less apparent in the tropics: they may be slower, more subtle, or more difficult to detect in the absence of widespread monitoring networks. They are certainly underway at higher latitudes.[74] Ecologically significant climate changes—large enough to drive major shifts in species distributions, alter ecological zonation patterns in unpredictable ways, and force widespread extinctions—are predicted for the Andes, as well as the rest of the tropics, if greenhouse gases continue to climb at current rates.[75]

In the *Essay on the Geography of Plants*, Alexander von Humboldt laid the foundation for the sciences that are telling us how our planet works, how our activities are affecting it, and how the workings of our planet, particularly

72. The scientific debate over the reality of human-induced climate change is long over. The 2007 Intergovernmental Panel on Climate Change (IPCC) report summarizes the scientific evidence. Some scientists now argue that the IPCC report is too conservative, particularly with respect to glacier melting, sea-level rise, and positive feedbacks (see Kerr 2007). Several excellent popular books are now available; those by Flannery (2005) and Cox (2005) are particularly recommended.

73. Andean glaciers are retreating from Colombia to Chile. This is particularly frightening because glacial meltwater is critical for water supply, agriculture, and power generation in the region, which experiences seasonal drought. See Coudrain et al. 2005, Bradley et al. 2006, and Thompson et al. 2006.

74. Poleward shifts in species ranges are documented by Parmesan and Yohe (2003) and Root et al. (2003). Recent increases in wildfire magnitude in the American West are related to warming (Westerling et al. 2006), as are insect-pest outbreaks that have killed hundreds of millions of trees (Breshears et al. 2005, Berg et al. 2006, Hicke et al. 2006, Taylor et al. 2006).

75. See Williams et al. 2007 and Williams and Jackson 2007 for a recent analysis. If CO_2 concentrations double beyond preindustrial levels, tropical regions across the globe will undergo climate changes sufficiently large to force major changes in regional vegetation physiognomy. The characteristic climates of the tropical Andes will vanish, with likely widespread disruption and extinction. We are not quite halfway to doubling as I write this, and will easily get there by mid-century at current rates. If CO_2 concentrations triple—which will happen by 2100 AD under a "business-as-usual" scenario— the entire globe will undergo climate change sufficient to drive biome-level ecological conversions. Likely development of novel climates—combinations of seasonal temperatures and precipitation with no modern counterparts—under global climate change renders precise prediction of ecological responses highly uncertain, virtually ensuring that we will face ecological surprises.

its climate, affect human activities and welfare in turn. Humboldt had great faith in human reason and judgment. We will need every scrap of human reason and judgment we can muster to avoid going over the brink of climate disaster in the coming years. We have the power today to determine what Chimborazo will look like in the future—whether it will still have snow, whether the species Humboldt described in his *Tableau physique* will still grow on its slopes, and whether water and other resources will be sufficient to sustain the societies living in its shadow. We will need Humboldt's science, and his humanism, to exercise that power wisely.

ESQUISSE DES PRINCIPALES HAUTEURS DES DEUX CONTINENS

FIGURE 1. Goethe's comparative profile of the Andes and the Alps. Europe ("the ancient continent") is to the left, and the Americas ("the new continent") is to the right. The vertical scale is in toises (T). Goethe cleverly packed many different mountain summits into the figure by arranging the highest ones in the background and lower ones in the foreground. The background summits are highest at the flanks and decreased towards the center of the profile. Mont-Blanc (the highest point in Europe) is on the left flank at 2,450 T, with the Jungfrau (2,145 T) to its right, followed successively by "Pic de Teneriffe" (Pico de Teide, 1,905 T), Mount Etna (with a smoke plume, 1,713 T), and Canigou (1,427 T). The foreground includes various low summits (e.g., Vesuvius, 615 T, spewing smoke in the lower left), mountain passes (e.g., St. Gothard, 1,065 T), and cities (Madrid at 307 T, Paris at 18 T). Other notable features include Gay-Lussac's 1804 balloon ascent above Paris, Saussure standing at the summit of Mont-Blanc, and lower snowline in the Swiss Alps.

The Chimborazo summit is at the right flank, with Humboldt standing on its slope at 3,032 T. A condor soars some 300 toises above Humboldt's head. Cotopaxi (2,952 T), with a smoke plume, is to the left of Humboldt. Lower snowline, the upper limit of lichens, and upper treeline are all depicted for equatorial South America. Various towns and villages are also portrayed (Micuipampa, Quito, Bogotá, Mexico City). Vegetation zonation is suggested by palm trees near sea level in the foreground; the limit of palms and bananas is delimited at 513 T.

The title reads "Sketch of the principal heights of the two continents Prepared by Mr. Goethe, private counsel to the Duke of Saxe Weimar After the work of Mr. Humboldt, published in 1807 under the title *Essay on the Geography of Plants*."

Goethe drafted this sketch in 1807 after reading the German version of the *Essai* (which Humboldt had dedicated to Goethe) and sent it Humboldt. He later prepared a more polished version, printed in 1813 in the *Allgemeinen Geographischen Ephemeriden*, a journal published in Weimar by Goethe's friend, Friedrich Justin Bertuch (1747–1822). Goethe modified that figure to produce the one portrayed here, adding more details (e.g., lower snowline). This figure was reprinted as an insert in the 1963 reprint of Humboldt's *Ideen zu einer Geographie der Pflanzen*.

FIGURE 2. Humboldt's 1817 diagram provides an elegant summary of vegetation, snowline, and temperature patterns along elevational and latitudinal gradients. The analogy between latitude and elevation is immediately apparent to the viewer. This figure is less ambitious and more economical than the Tableau physique and conveys a great deal of information clearly and quickly.

The left-hand mountain is Chimborazo (equatorial), the middle mountain is Mont-Perdu, flanked by Mont-Blanc (temperate), and the right-hand mountain is Sulitelma in Lapland (subarctic). The lower elevation of snowline, the upper limit of treeline, and various vegetation zones all descend as one moves poleward. From Humboldt's 1817 De Distributione Geographica Plantarum.

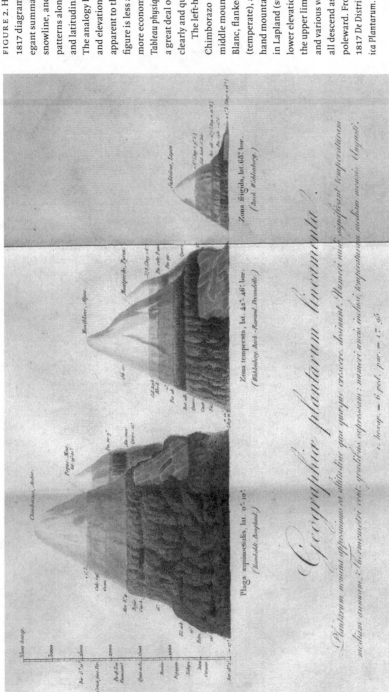

Geographiæ plantarum lineamenta.

FIGURE 3. This topographic profile of Chimborazo was published in 1826 and was distributed with Humboldt's Personal Narrative. The figure is more modest than the 1807 Tableau physique, confining itself to the plants arrayed along the slopes of the mountain (left-hand side) and various elevational features of interest. It is also a more literal representation than the Tableau physique, aimed at conveying a realistic sense of the topographic and physiographic context of Chimborazo. The margins include elevations of various features for comparison (Cotopaxi, Great St. Bernard Pass, Vesuvius, Quito), some temperature information (on the left margin), and the lower limit of snow in Mexico, the Himalayas, and the Alps. By the time this figure was prepared, the Himalayas were recognized as being higher than the Andes; the right-hand margin shows estimated heights of Himalayan peaks (> 4,000 T), well above Gay-Lussac's balloon.

FIGURE 4. Detail from the 1826 Chimborazo profile, showing the plant names arrayed along the slopes. In this version Humboldt appears to have restricted himself to plants of the immediate vicinity of Chimborazo; coastal and other plant names are not included, nor are the vegetation zones. Many names are abbreviated.

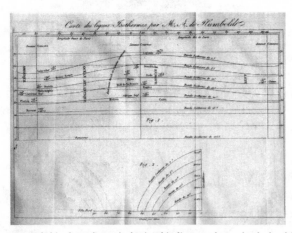

FIGURE 5. Humboldt pioneered the use of isothermal lines in these 1817 figures. The upper panel portrays the Northern Hemisphere from the equator to 70°, and from eastern North America to Europe and Asia. Humboldt's isothermal lines are relatively flat at low latitudes but become sinuous starting at 30°N, with concavities in eastern North America and convexities in Europe. The lower diagram shows north latitude on the horizontal axis and altitude on the vertical axis. This diagram shows clearly the altitude/latitude analogy for temperature, and the progressive erosion of higher-temperature isotherms with increasing latitude. From Humboldt 1817.

FIGURE 6. Friedrich Georg Weitsch's 1810 painting of Humboldt (standing) and Bonpland (sitting at lower right) in the Andes. The human figures are dwarfed by the Chimborazo massif in the background (left). Cotopaxi is in the background directly behind Humboldt and Bonpland. Weitsch prepared the painting from field sketches provided by Humboldt. From Stiftung Preussische Schlosser und Garten Berlin—Brandenburg.

FIGURE 7. Julius Schrader's portrait of Humboldt, painted in the months before Humboldt's death. Chimborazo and Cotopaxi loom in the background, unchanged in the half-century between the paintings. This is not surprising, since neither Schrader nor Weitsch visited the Andes, basing their images of Chimborazo on Humboldt's 1802 sketches. Nonetheless, the paintings together convey a sense of a timeless, unchanging mountain landscape, with only the human figure affected by the passing of time. From the Metropolitan Museum of Art.

Translator's Note

This translation has been guided by two overall principles: remaining faithful to Humboldt's tone and ease of reading for the modern audience. Humboldt was a scientific thinker, writer, and explorer, doing scientific work on the terrain and preparing his writing at the same time. The prose in this work is objective and unemotional, and although he certainly wants to appeal to the public, he prefers letting the facts speak for themselves. He is not striving for the picturesque, but, as an Enlightenment scientist, he is intent on making knowledge widely available and on persuading his readers of its value. If very occasionally an emotion pierces through—and there is no doubt he was moved by the grand spectacle of nature he saw—he avoids poetic language, alluding to his emotion more than showing it. This translation has sought to remain faithful to this intention and this tone of voice.

Making the text easy to read and transparent to the modern reader has entailed making some choices, some of which reflect Humboldt's own usage, some of which do not. Where I could keep some of the flavor of the text, I did so as long as clarity could be preserved. Latin names have been left as in the original, as has the punctuation, so that there are multiple sentences separated by semi-colons or colons as Humboldt wrote them. References to the older measures as well as to the newer metric ones have been kept. Some changes have been made for ease of reading. Spelling (for both English and French quotations) has been modernized, and for numbers I have used numerals rather than words for clarity. Instead of the modern way of indicating latitude, I have kept the words "boreal" and "austral," but for areas of countries or continents, I have generally used "northern" and "southern." A few words could not be translated and have been left in French in square brackets. Place names pose a special difficulty, and some of Humboldt's own spellings vary. For as many names as possible, Spanish names have been spelled as they are today in any readily available atlas.

Sylvie Romanowski

Essay on the Geography of Plants

Alexander von Humboldt and Aimé Bonpland

Essay on the Geography of Plants

Translated by Sylvie Romanowski

The Voyage of Humboldt and Bonpland

FIRST PART

General Physics and Historical Account of the Voyage

THE VOYAGE

OF

HUMBOLDT AND BONPLAND

FIRST PART

GENERAL PHYSICS AND HISTORICAL ACCOUNT OF THE VOYAGE

FIRST VOLUME

Containing an Essay on the Geography of Plants,
together with a Physical Tableau of the Equinoctial
Regions, as an Introduction to this Work

With one Plate

PARIS

Fr. Schœll, Bookseller, 19 Maçons-Sorbonne Street
and Tübingen, J. G. Cotta, Bookseller

1807

ESSAY
ON THE GEOGRAPHY OF PLANTS
WITH
A PHYSICAL TABLEAU OF THE EQUINOCTIAL REGIONS

Based on measurements made from the tenth degree
of boreal latitude to the tenth degree of austral latitude in
the years 1799, 1800, 1801, 1802, and 1803

By Al. de Humboldt and A. Bonpland

Written by Al. de Humboldt

With one Plate

PARIS

Fr. Schœll, Bookseller, 19 Maçons-Sorbonne Street
and Tübingen, J. G. Cotta, Bookseller

1807

For Messrs
Antoine Laurent de Jussieu
and
René Desfontaines

Professors at the Museum of Natural History
Members of the National Institute, etc.

Having left Europe five years ago, and having traveled through some countries never seen before by naturalists, I might have hastened to publish earlier an abridged account of my travels in the tropics and the various phenomena that I studied one after another. Perhaps I might have flattered myself that such a publication would be received by the public, some of whom showed warm interest for my personal well-being as well as for the success of my expedition. But I thought that before talking about myself and the obstacles that I had to overcome during my work, I should draw the physicists' attention to the broader phenomena exhibited by nature in the regions through which I have traveled. What I offer here is a comprehensive view of these phenomena. This essay gives the results of the observations that are developed in greater detail in other works that I am preparing for publication.

Here I bring together all the physical phenomena that one can observe both on the surface of the earth and in the surrounding atmosphere. The physicist who is acquainted with the current state of science, especially that of meteorology, will not be surprised to see that so many topics are discussed in so few pages. If I could have worked on this book for a longer time, it would have been even shorter; for a tableau must contain only the general physical qualities and results that are certain and able to be expressed in exact numbers.

I conceived of this book during my earliest years. I gave a first sketch of a *Geography of Plants* in 1790 to Cook's famous colleague, Mr. Georges Forster, with whom I had close ties of friendship and gratefulness. My later research in various areas of physics helped me reach a wider understanding of my initial ideas. My voyage in the tropics furnished me with precious materials for the physical history of the earth. I wrote the major part of this work in the very presence of the objects I was going to describe, at the foot of the Chimborazo, on the coasts of the South Sea. I thought it best to keep the title *Essay on the Geography of Plants*, because any less modest title might have revealed its imperfections and rendered it less worthy of the public's indulgence.

I must plead for this indulgence especially for the style: having been obliged to use several languages that just like French are not my own, I cannot hope to express myself in as pure a style as might be expected if I used my own language.

The tableau that I offer here is the result of my own observations and those of Mr. Bonpland. We are linked by the closest of friendships, having worked together for six years, and sharing the same troubles to which the traveler is necessarily exposed in uncivilized lands: for these reasons we decided that all the works resulting from our expedition would be published under our two names.

In revising these works, as I have been doing since my return from Philadelphia, I have had recourse to famous men who honor me with their kindness. Mr. Laplace, whose fame is beyond my ability to praise him, was kind enough to show his interest, thus flattering greatly the work I brought back as well as the work I believed I had to write after my return from Europe. His genius illuminates and enlivens, so to speak, everything that surrounds him, so that his kindness has become as useful to me as to all the young people who come to him.

It is a great pleasure to bring him the tribute of my admiration and my gratitude, and it is no less a sacred duty to fulfill the duties of friendship. Mr. Biot was kind enough to honor me with his advice in the writing of this book. Since he is as wise a physicist as he is profound a geodesist, my relationship with him has become for me a fertile source of learning: in spite of the large number of his engagements, he was kind enough to calculate the tables of the horizontal refractions and of the extinction of light which are appended to my tableau.

The facts that I state on the history of fruit trees come from the work of Mr. Sickler, containing both great erudition and very philosophical views, which are rarely found together.

Mr. Decandolle furnished me with interesting materials relating to the Geography of Plants in the upper Alps; Mr. Ramond gave me materials on the flora of the Pyrenees; I found others in the classic works of Mr. Wildenow. It was important to compare the phenomena of equinoctial vegetation with those found in our European soil. Mr. Delambre was kind enough to enrich my tableau with several measurements of altitudes which have never been published. A large number of my barometric observations were calculated by Mr. Prony according to Mr. Laplace's formula, taking into account the influence of gravity. This worthy scientist took his kindness so far as to have more than four hundred of my measurements of altitude calculated under his personal supervision.

At this moment, I am writing about the astronomical observations that I made during my expedition, a great number of which were presented at the Bureau des Longitudes in order to determine their exactitude. Before this is done, it would be unwise to publish the maps I drew up of the interior of the continents, or even the account of my travels, since the position of the sites and their altitude influence all the phenomena in the regions I visited. I am hopeful that especially the observations that I made during my navigations on the Orinoco, the Casiquiare, and the Río Negro will be of interest to those who study the geography of South America. Despite Father Caulin's precise description of the Casiquiare, modern geographers have expressed new doubts on the link between the Orinoco and the Amazon Rivers. Having worked in this location, I did not expect that I would encounter bitter opposition[1] to my having found in nature very different river flows than those indicated by the Cruz map; but it is the fate of travelers that they may displease others when they observe facts that are contrary to received ideas.

After I have written the book on astronomy, I will be able to proceed faster with my other works; only after having published the results of my latest voyage will I undertake a new project, which may shed much light on meteorology and magnetic phenomena.

I cannot publish this essay, the first fruit of my research, without expressing my profound and respectful gratitude to the government that honored me with its generous protection during my travels: enjoying a permission never granted before to any individual, I lived for five years in the midst of an honest and loyal nation, and never encountered in the Spanish colonies any obstacles other than those of physical nature itself. The memory of the government's kindness will remain permanently in my soul, as will the interest and affection on the part of people from all classes who honored me during my stay in the two Americas.

Alex. de Humboldt

1. *Géographie moderne* [Modern Geography], by Pinkerton, trans. Walkenaer; vol. 6, pp. 174–77.

Essay on the Geography of Plants[2]

Botanists usually direct their research towards objects that encompass only a very small part of their science. They are concerned almost exclusively with the discovery of new species of plants, the study of their external structure, their distinguishing characteristics, and the analogies that group them together into classes and families.

This knowledge of the forms which make up organized beings is no doubt the principal basis for descriptive natural history. It must be regarded as indispensable for the advancement of the sciences that concern the medical properties of plants, their cultivation, or their applications in the arts; even if this knowledge is worthy of occupying a great number of botanists, even if it can be considered from a philosophical point of view, it is no less important to understand the Geography of Plants, a science that up to now exists in name only, and yet is an essential part of general physics.

This is the science that concerns itself with plants in their local association in the various climates. This science, as vast as its object, paints with a broad brush the immense space occupied by plants, from the regions of perpetual snows to the bottom of the ocean, and into the very interior of the earth, where there subsist in obscure caves some cryptogams that are as little known as the insects feeding upon them.

The upper limit of vegetation varies, like that of perpetual snows, according to the distance of the location from the poles or the slant of the sun's rays. We do not know the lower limit of vegetation: but some precise observations carried out on subterranean vegetation in both hemispheres prove that the interior of the earth supports life wherever organic seeds have found a place adequate for their development and the appropriate sustenance for their organism. The rocky and icy peaks above the clouds, barely discernible to the eye, are covered only with mosses and lichenous plants. Similar cryptogams, sometimes pale, sometimes colorful, branch out on the roofs of mines and underground caves. Thus the opposite limits of plant life produce beings with a similar structure and a physiology equally unknown to us.

2. Read in the Physics and Mathematics Class at the Institut National, 17th of Nivôse Year XIII [January 7, 1805].

The geography of plants does not merely categorize plants according to the various zones and altitudes where they are found; it does not consider them merely in relation to the conditions of atmospheric pressure, temperature, humidity, and electrical tension in which they live; it can discern, just as in animals, two classes having a very different kind of life, and, so to speak, very different habits.

One class of plants grows in an isolated and sparse fashion: such are, in Europe, *Solanum dulcamara*, *Lychnis dioica*, *Polygonum bistorta*, *Anthericum liliago*, *Crataegus aria*, *Weissia paludosa*, *Polytrichum piliferum*, *Fucus saccharinus*, *Clavaria pistillaris*, *Agaricus procerus*; in the tropics, *Theophrasta americana*, *Lysianthus longifolius*, *Cinchona*, *Hevea*. Another class of plants live in an organized society like the ants and the bees, and occupy immense terrains from which they exclude any heterogeneous plants. Such plants are strawberry plants (*Fragaria vesca*), bilberry plants (*Vaccinium myrtillus*), *Polygonum aviculare*, *Cyperus fuscus*, *Aira canescens*, *Pinus sylvestris*, *Sesuvium portulacastrum*, *Rhizophora mangle*, *Crotum argenteum*, *Convolvulus brasiliensis*, *Brathys juniperina*, *Escallonia myrtilloides*, *Bromelia karatas*, *Sphagnum palustre*, *Polytrichum commune*, *Fucus natans*, *Sphaeria digatata*, the lichen *Haematomma*, *Cladonia paschalis*, *Thelephora hirsuta*.

These socially organized plants are more common in temperate zones than in the tropics, where the vegetation is less uniform and thereby more picturesque. From the banks of the Orinoco to those of the Amazon and the Ucayali Rivers, for over five hundred leagues the entire surface of the land is covered with thick forests; and if the rivers did not break up this continuum, the monkeys who are almost the sole inhabitants of these solitary places could travel from the boreal to the austral hemisphere by leaping from branch to branch. But these immense forests do not offer a uniform kind of socially organized plant life; each part produces different kinds. Here one finds mimosas, *Psychotria*, or melastomes, there one finds laurel, Caesalpinaceae, *Ficus*, *Carolinea*, and *Hevea*, that interlace their branches: no one plant dominates over the others. This is not the case in the tropical region close to New Mexico and in Canada. From the 17th degree to the 22nd degree of latitude, the entire Anahuac region, the entire plateau lying from 1,500 to 3,000 meters above sea level, is covered with oaks and with a kind of pine tree that is close to *Pinus strobus*. On the eastern slope of the Cordillera, in the valleys of Jalapa, one finds a large forest of liquidambars: the soil, the vegetation, and the climate are like those of temperate regions; this is not observed in any other part of South America at a similar altitude.

The reason for this phenomenon seems to depend on the structure of the American continent. This continent widens out towards the north pole and continues in this direction much more than the European continent; this makes the Mexican climate much colder than it would be according to its latitude and its height above sea level. The plants in Canada and those in the more northerly regions spread towards the south, and the volcanic mountains in Mexico are covered with the same pine trees that would seem to belong only to the sources of the Gila and the Missouri.

In Europe, on the contrary, the great catastrophe which opened up the Strait of Gibraltar and fashioned the Mediterranean seabed prevented the plants in Africa from spreading since then into northern Europe: thus very few such plants are found north of the Pyrenees. But the oaks that crown the heights of the valley of Tenochtitlan are identical to the species at the 45th degree, and if a painter traveled through this tropical region to study its vegetation, he would not find the beauty and variety found in equinoctial plants. In the parallel of Jamaica, he would find forests of oaks, pines, *Cupressus disticha* and *Arbutus madronno*; these forests are very similar and just as monotonous as the social plants in Canada, Europe, and northern Asia.

It would be interesting to show on botanical maps the areas where these groupings of similar species of plants live. These maps would show long bands, whose irresistible extension causes the population of states to decrease, the nations to be separated, and creates stronger obstacles to communication than do mountains and seas. Heath, an association of *Erica vulgaris*, *Erica tetralix*, the lichens *Icmadophila* and *Haematomma*, spreads from the most northerly extremity of Jutland, through Holstein and Luneburg, up to the 52nd degree of latitude. From there it spreads towards the west, through the granitic sands of Munster and Breda, and to the Ocean coasts.

For a long stretch of centuries, these plants have made the soil sterile and have dominated these regions completely: despite man's efforts and his fight against an almost unconquerable nature, he has been able to claim only small areas for agriculture. These cultivated fields, the only fruits of hard work to be beneficial for humanity, are like small islands in the middle of the heath; they recall to the traveler's imagination those oases in Libya, so green and fresh in contrast to the surrounding desert sands.

A moss common to tropical and European marshes, *Sphagnum palustre*, used to cover a large part of Germania. This very moss made large areas of land uninhabitable for the nomadic peoples described by Tacitus. A geological feature explains this phenomenon. The oldest peat bogs, which have a mélange of sodium chloride and marine shells, arose from ulvae and *Fucus*; the most recent ones and the most widespread, on the contrary, arise from

Sphagnum and Mnium serpillifolium; their existence thus proves how abundant these cryptogams once were on the earth. In cutting down the forests, agricultural peoples caused a decrease in the humidity of the climate; the marshes dried up, and useful plant life spread little by little over the plains that used to be occupied exclusively by cryptogams unfavorable to agriculture.

Even though the phenomenon of socially organized plant life seems to belong primarily in temperate zones, the tropics can also furnish several examples of this. Along the top of the long Andean chain of mountains, at an elevation of 3,000 meters, one can find Brathis juniperina, Jarava (a type of grass close to Papporophorum), Escallonia myrtilloides, several species of Molina, and especially Tourrettia, whose marrow provides a food that the impoverished Indian sometimes has to wrestle away from the bears. In the plains separating the Amazon River from the Chinchipe, one can find together Croton argentum, Bougainvillea, and Godoya; just as in the savannas of the Orinoco, one can find the palm tree Mauritia, herbaceous sensitive plants, and Kyllingia. In the kingdom of New Grenada, the Bambusa and the Heliconia occur in long uniform bands that are not interrupted by other plants: but these associations of the same species of plants are less consistently extensive, less numerous, than in temperate climates.

In order to determine the ancient link between neighboring continents, geology bases itself on the analogy of the coastal structures, on the ocean beds, and on the types of animals living there. The geography of plants can furnish precious materials for this kind of research: up to a point, it can show how islands that were previously linked are now separated; it can show that the separation of Africa from South America occurred before the development of organized forms of beings. This same science can determine which plants are common to east Asia and to the coasts of Mexico and California; it can determine whether there are any plants which live in every zone and in every elevation above sea level. The geography of plants can assist us in going back with some degree of certainty to the initial state of the earth: it can determine if, after the initial retreat of the waters which were abundant and agitated as attested by rocks filled with sea shells, the entire surface of the globe was covered at once with various plants, or whether, as traditional accounts of various peoples say, the earth, now stilled, produced plants only in one area, from which they were transported by sea currents in a progressive march to far-flung regions over the course of centuries.

This same science can examine whether among the immense variety of plant forms one can recognize some primitive forms, and whether the diversity of species can be considered to be an effect of the degeneration that over time transformed accidental varieties into permanent ones.

If I may draw some general conclusions from the phenomena I observed in both hemispheres, it seems to me that the seeds of cryptogams are the only ones that nature develops spontaneously in all climates. Dicranium scoparium and Polytrichum commune, Verrucaria sanguinea and Verrucaria limitata of Scopoli are found in all latitudes, in Europe as well as at the equator, and not only in the tallest mountain ranges, but even at sea level, wherever there is shade and humidity.

On the banks of the Magdalena, between Honda and the Egyptiaca, in a plain where the centigrade thermometer hovers almost constantly between 28 and 30 degrees, at the base of Macrocnemum and Ochroma trunks, mosses form lawns as green and as beautiful as those in Norway. Other travelers did affirm that cryptogams were very rare in the tropics, but this assertion was no doubt based on the fact that they went only near dry coasts or in small cultivated areas, without going far enough into the interior of the continents. Lichenous plants are found in every latitude: their shape seems as independent of the influence of the climates as is the kind of rocks they live on.

We do not yet know of any phanerogamous plant whose organs are flexible enough to adapt to all zones and all altitudes. In vain has it been asserted that Alsine media, Fragaria vesca, and Solanum nigrum enjoyed this advantage, a capacity shared only by man and a few mammals that he maintains around him. The strawberry of the United States and Canada is different from the one in Europe. Mr. Bonpland and I thought we had discovered a few of these plants in the Andean Cordillera while passing from the valley of the Magdalena to the Cauca valley through the snows of Quindiu. The solitary nature of these forests composed of styrax, passiflora trees and wax palm trees, the lack of agriculture in the environs, and other circumstances, seemed to exclude the possibility that these strawberry plants were brought here by man or by birds; but perhaps had we seen this plant in bloom, we would have found specific differences with Fragaria vesca, just as there are very slight nuances between Fragaria elatior and Fragaria virginiana: in any case, in the five years that we studied plants in both hemispheres, we never encountered any European plant produced spontaneously by the South American soil. We must be content with believing that Alsine media, Solanum nigrum, Sonchus oleraceus, Apium graveolens, and Portulaca oleracea are plants that, like the peoples of the races in the Caucasus, are very widespread in the northern part of the Old Continent. We know so little about what is produced in the interior of the lands that we must abstain from any general conclusion: we might fall into the error of those geologists who construct the entire earth according the shape of the nearest hills surrounding them.

In order to solve the great problem of the migration of plants, the geography of plants delves into the interior of the earth: it looks at the ancient monuments that nature has left behind in petrifications, in wood fossils, and in coal strata that are the tomb of the initial plant life of our planet. This science can discover petrified fruits in India, and palm trees, tree ferns, scitaminales, and tropical bamboos buried in the ice-covered lands in the north; it can consider whether these equinoctial productions, as well as the bones of elephants, tapirs, crocodiles, and marsupials which have been found recently in Europe, can have been carried to these temperate zones by the strength of currents in a submerged world, or whether these same climates formerly sustained the palm trees and the tapir, the crocodile and the bamboo. One can lean towards the latter supposition, when one considers the local circumstances that occur in Indian petrifications. But can one accept that such great changes in the atmosphere's temperature occurred without having recourse to a displacement of planets, or a change in the earth's axis—unlikely suppositions given our current knowledge of astronomy? Considering that the most striking geological phenomena show that the whole crust of the earth was once in a liquid state; that the stratification and differentiation of rocks indicate that the formation of mountains and the crystallization of the great land masses around a common nucleus did not happen at the same time on the whole surface of the earth; then one can conceive that the passage from liquid to solid states must have liberated an immense quantity of heat, and thereby increased the temperature of a region independently of solar heat: but would this local increase in temperature have lasted long enough to explain the nature of these phenomena?

Changes observed in the light of celestial bodies have led to the suspicion that the one which is at the center of our system undergoes similar variations. Would an increase in the intensity of the sun's rays have spread in certain periods tropical heat to the zones near the poles? Do such variations, which would enable equinoctial plants, elephants, and tapirs to live in Lapland, occur periodically? or are they the result of some temporary disturbances in our planetary system?

These are topics by which the geography of plants is related to geology. By shedding light on the prehistory of our planet, it offers to the human imagination a field to cultivate that is as rich as it is interesting.

Though plants are very analogous to animals regarding the irritability of their fibers and the stimuli that excite them, they are essentially different from animals as regards their mobility. Most animals leave their mother only when they reach adulthood. On the contrary, plants, being fixed in the soil after their development, can move only when still contained in the egg, whose structure favors mobility. But winds, currents, and birds are not the only ones that help plants migrate; the primary factor is man.

When man abandons the wandering life, he gathers around him the plants and animals that can feed and clothe him. This passage from nomadic life to agriculture occurs late in the northern peoples. In equinoctial regions, between the Orinoco and the Amazon, the density of the forest prevents the savage from living solely by means of hunting: he has to grow a few plants, some Jatropha, banana trees, and Solanum, for subsistence. Peach trees, the fruit of palm trees, and some small farms (if such a small group of cultivated plants can be so called), these are the basis for the South American Indians' food. The character of the savage is modified everywhere by the nature of the climate and the soil where he lives. These modifications alone accounted for the differences between the first inhabitants of Greece and the Bedouin herders, and between these and the Indians in Canada.

A few plants grown in gardens and farms since times immemorial accompanied man from one end of the planet to the other. In Europe, the Greeks took with them vines, the Romans, wheat, and the Arabs, cotton. In America, the Toltecs carried maize with them: potatoes and quinoa are found wherever went the inhabitants of the ancient Cundinamarca. The migration of these plants is obvious; but the land of their origin is as unknown as the origin of the various human races found all over the earth in the remotest times according to traditional accounts. South and east of the Caspian sea, on the banks of the Amu Darya, in the ancient Colchis and especially in the province of Kurdistan, whose mountains are perpetually covered with snow and are hence over 3,000 meters, there the land is covered with lemon trees, pomegranate trees, cherry trees, pear trees, and all the fruit trees that we grow in our gardens. We do not know if these trees are native to those regions or whether they were formerly cultivated and became wild, and thereby attest to the antiquity of their cultivation in these regions. Situated between the Euphrates and the Indus, between the Caspian Sea and the Persian Gulf, these fertile lands gave Europe its most precious products. Persia gave us the walnut tree and the peach tree; Armenia, the apricot tree; Asia Minor, the cherry tree and the chestnut tree; Syria, the fig tree, the pear tree, the pomegranate tree, the olive tree, the plum tree, and the mulberry tree. In Cato's time, the Romans did not yet know the cherry, the peach, or the mulberry.

Hesiod and Homer already mention the cultivation of the olive tree in Greece and in the islands of the Archipelago. In the reign of Tarquinius the Ancient, this tree did not yet exist in Italy, Spain, or Africa. During Appius Claudius's consulate, oil was still very scarce in Rome; but by Pliny's time the olive tree had already spread to France and Spain. The vines that we grow today are not indigenous to Europe: they grow wild on the coasts of the Caspian Sea, in Armenia, and in Caramania. From Asia they spread to Greece and from there to Sicily. The Phoceans carried them to southern France: the Romans planted them on the banks of the Rhine. The *Vites* species growing wild in North America that gave its name (Vinland) to the first part of the land discovered by the Europeans is very different from our *Vitis vinifera*.

A cherry tree laden with fruit adorned Lucullus's triumph; it was the first tree of its kind ever seen in Italy. The dictator took it from the province of Pontus when he was victorious over Mithridates. In less than a century, the cherry tree became widespread in France, Germany, and England. So does man change at will the surface of the earth, gathering around him plants coming from the most distant climates. In the European colony of the two Indies, one small cultivated plot may contain coffee from Arabia, sugar cane from China, indigo from Africa, and a host of other plants belonging to both hemispheres. This variety of plants is interesting in that it recalls to one's imagination a series of events that caused the human race to spread over the entire surface of the earth and to appropriate all its productions.

In this manner, man, being restless and industrious, traveled in all the earth's regions and thereby forced a certain number of plants to live under many climates and in many altitudes; but the domination that he exercised over these organized beings did not modify their primitive structure. Grown in Chile at an altitude of 3,600 meters (1,936 toises), potatoes have the same blossoms as the ones introduced in the Siberian plains. Barley that was used to feed Achilles' horses is no doubt the same as the one grown today. The characteristic forms of the plants and animals that occupy the earth today do not seem to have undergone any changes since the remotest times. The ibis buried in the Egyptian catacombs, a bird that goes back almost to the time of the pyramids, is identical to the one fishing today on the banks of the Nile; this identity shows that the huge amounts of animal fossils found in the interior of the earth do not belong to varieties of current species, but to a very different order of things from ours, and too ancient for any of the traditions to remember.

Having introduced new plants, man preferred to cultivate these and made them dominant over the indigenous ones; but this preponderance of new plants, which makes the European cultures seem so monotonous and hopelessly dull to the botanist during his excursions, prevails only in a small part of the planet where civilization perfected itself and where consequently population increased the most. In the lands near the equator, man is too weak to tame a vegetation that hides the ground from view and leaves only the ocean and the rivers to be free. There nature demonstrates its wildness and majesty that render impossible all efforts of cultivation.

The origin and the first homeland of the plants most useful to man and living with him since the remotest times is as impenetrable a secret as that of the land of origin of domestic animals. We do not know the homeland of the grasses that are the primary subsistence for the peoples of the Mongolian race and of the Caucasus; we do not know in which region the cereals spontaneously arose, such as wheat, barley, oats, and rye. This latter species of grasses does not seem to have been grown even by the Romans. Some have maintained that these plants were found to grow wild, barley on the banks of the Samara in the land of the Tartars, *Triticum spelta* in Armenia, rye in Crete, wheat in Bashkiria in Asia: but these facts do not seem to be very well ascertained; it is too easy to confuse plants growing spontaneously with plants that fled from man's domination and regained their original freedom. Birds eating the seeds of cereals can easily disseminate them throughout the forests. Plants constituting the native wealth of all the tropics' inhabitants—the banana tree, *Carica papaya*, *Jatropha manihot*, and maize— have never been found in a wild state. I have seen a few such plants on the banks of the Casiquiare and the Río Negro: but the savages of this region, both melancholy and distrustful, cultivate small plots of land in the most solitary places; they abandon them a short while later, and the plants they left behind thus seem indigenous to the soil where they grow. The potato, so beneficial in feeding large populations in the most sterile regions of Europe, shares the same characteristics as the banana tree, maize, and wheat. However much research I conducted on location, I never learned that any traveler found it growing wild either on the summit of the Peruvian Cordillera or in the kingdom of New Grenada, where this plant is grown together with the *Chenopodium quinoa*.

Such are the characteristics found in agriculture and its various products, varying with their latitude or with the origin and needs of the peoples. The impact of food that can be more or less stimulating to the character and strength of the passions, the history of navigations and wars carried out over the products of the plant kingdom; such are the factors that link the geography of plants to the political and intellectual history of mankind.

These relationships would be no doubt sufficient to show how extended is the science which I am attempting to outline here; but the man who is sensitive to the beauty of nature will also find here the explanation for the influence exerted by nature on the peoples' taste and imagination. He will delight in examining what is called the character of vegetation, and the variety of effects it causes in the soul of the observer. These considerations are important in that they are closely related to the means by which imitative arts and descriptive poetry can affect us. Merely looking at nature, at its fields and forests, causes a pleasure that is essentially different from the impression given by studying the specific structure of an organized being. In the latter, the details interest us and excite our curiosity; in the former, the large picture, the ensemble, excites our imagination. How different is the aspect of a vast prairie surrounded with a few clumps of trees from that of a dense, dark forest of oaks and pines? How different are the forests of temperate zones from those of the equator where the naked and thin trunks of palm trees soar above the flowering mahogany trees and resemble majestic porticos? What is the intellectual cause of these feelings? Are they produced by nature, by the large size of these ensembles, by the outline of their shapes, or by the plants' posture? How does this posture, this more or less rich or cheerful aspect of nature influence the habits and sensibilities of peoples? What is the character of tropical vegetation? What features distinguish the African plants from those of the New World? What are the analogies in shape that link the alpine plants of the Andes with those of the high peaks of the Pyrenees? These questions have hardly been debated till now and are without a doubt worthy of the physicist's attention.

Among the variety of plants covering the surface of our planet, one can easily distinguish certain general forms under which the others can be subsumed, and which can be arranged into families or groups that are more or less analogous to each other. I will name only fifteen of these groups, whose aspect is most interesting to the painter of landscapes: (1) the scitaminales form (*Musa, Heliconia, Strelitzia*); (2) the palms; (3) the tree-ferns; (4) the form of *Arum, Pothos,* and *Dracontium*; (5) the pines (*Taxus, Pinus*); (6) heaths *folia acerosa*; (7) the tamarind form (*Mimosa, Gleditsia, Porlieria*); (8) the Malvaceae form (*Sterculia, Hibiscus, Ochroma, Cavanillesia*); (9) the lianas (*Vitis, Paullinia*); (10) the orchids (*Epidendrum, Seratis, Paullinia*); (11) the prickly-pears (*Cactus*);

(12) the casuarines and *Equisetum*; (13) the grasses; (14) the mosses; (15) lastly, the lichens.

These divisions based on physiognomy have almost nothing in common with those made by botanists who have hitherto classified them according to very different principles. Only the outlines characterizing the aspect of vegetation and the similarities of impressions are used by the person contemplating nature, whereas descriptive botany classifies plants according to the resemblance of their smallest but most essential parts, those relating to fructification. An artist of distinction would find it worthwhile to study, not in greenhouses or in botany books but in nature itself, the physiognomy of the plant groups that I have enumerated. How very interesting for a tableau would be the antique trunk of a palm tree, with its variegated leaves swaying above a group of *Heliconia* and banana trees! How interesting would be the picturesque contrast of a tree-fern surrounded by Mexican oaks!

The absolute beauty of these shapes, their harmony, and the contrast arising from their being together, all this makes what is called the character of nature in various regions. Some shapes, indeed, and the most beautiful ones (the scitaminales, palm trees, and bamboos), are missing entirely from our temperate regions; others, such as trees with pinnate leaves, are very rare and less elegant. Arborescent species are very few, not so tall, and less laden with flowers pleasing to the eye. Thus the number of social plants, as described earlier, and man's cultivation make the aspect of the land more monotonous. In the tropics, on the contrary, nature delighted in combining every possible shape. The shapes of pines seem to be missing at first glance; but in the Andes of Quindiu, in the temperate forests of Oxa and in Mexico, there are cypress, pines, and junipers.

In general near the equator plant shapes are more majestic and more imposing; the leaves shine more brilliantly, and the texture of the parenchyma is looser and more succulent. The tallest trees are consistently adorned with flowers that are more beautiful, bigger, and more perfumed than those of herbaceous plants found in temperate zones. The singed bark of their ancient trunks makes a most agreeable contrast with the tender green leaves of the lianas, the pothos, and especially the orchids, whose flowers imitate the shape and the feathers of the birds feeding upon their nectar. However, in the tropics, we almost never see the breadth of the green prairies that lie along the rivers in northern countries; we almost never feel there the sweet feeling of springtime awakening the vegetation. Nature, beneficial to all beings, apportioned its specific gifts for each region. There are fibers that are more or less loose, plant colors that are more or less vivid according to the chemical nature of their elements and the strength of the sun's stimulation: these are some of the causes of the varieties of plants that give each zone of the planet its particular character. The great elevation of the lands near the equator presents to the inhabitants of the tropics a curious spectacle where the plants have the same shapes as those of Europe.

The valleys in the Andes are adorned with banana and palm trees; at a higher elevation one can find a beneficial tree whose bark is the fastest and healthiest fever reducer. In this temperate elevation where quinquinas are found, and in higher elevations where there are *Escallonia*, there one finds oaks, pines, *Berberis*, *Alnus*, *Rubus*, and a host of genera that we think belong only in the northern lands. Thus the inhabitants of equinoctial regions know all the species that nature placed around them: the earth offers to their eyes a spectacle as varied as the azure vault of the heavens which hides none of its constellations from view.

The Europeans do not enjoy such a spectacle. The frail plants that people, out of love of science or refined luxury, grow in their hothouses are mere shadows of the majestic equinoctial plants; many of these shapes will remain forever unknown to the Europeans; but the richness and perfection of the languages, the imagination and sensitivity of the poets and the painters give some compensation in Europe. The varied spectacles of the equinoctial regions are given to us by the imitative arts. In Europe, a man isolated on an arid coast can enjoy in thought the picture of faraway regions: if his soul is sensitive to works of art, if he is educated enough in spirit to embrace the broad concepts of general physics, he can, in his utter solitude and without leaving his home, appropriate everything that the intrepid naturalist has discovered in the heavens and the oceans, in the subterranean grottos, or on the highest icy peaks. This is no doubt how enlightenment and civilization have the greatest impact on our individual happiness, by allowing us to live in the past as well as the present, by bringing to us everything produced by nature in its various climates, and by allowing us to communicate with all the peoples of the earth. Sustained by previous discoveries, we can go forth into the future, and by foreseeing the consequences of phenomena, we can understand once and for all the laws to which nature subjected itself. In the midst of this research, we can achieve an intellectual pleasure, a moral freedom that fortifies us against the blows of fate and which no external power can ever reach.

Physical Tableau of Equatorial Regions

Based on measurements and observations performed on location, from the tenth degree of boreal latitude to the tenth degree of austral latitude in the years 1799, 1800, 1801, 1802, and 1803.

When one ascends from sea level to the peaks of high mountains, one can see a gradual change in the appearance of the land and in the various physical phenomena in the atmosphere. The plants in the plains are gradually replaced by very different ones: woody plants decrease little by little and are replaced by herbaceous and alpine plants; higher still, one finds only grasses and cryptogams. Rocks are covered with a few lichens, even in the regions of perpetual snow. As the appearance of the vegetation changes, so does the form of the animals: the mammals living in the woods, the birds flying in the air, even the insects gnawing at the roots in the soil are all different according to the elevation of the land.

By looking carefully at the nature of rocks of the earth's crust, the observer can also see changes in them as he climbs above sea level. Sometimes, the more recent formations covering the granite in the plains reach only a certain altitude; and near the mountain peaks this same primitive rock reappears that is the basis for all the others, and which constitutes the interior of our planet, so far as our feeble endeavors have allowed us to penetrate it.[3] Sometimes, this granitic rock remains hidden under other more recent formations. Peaks over 4,000 meters (2,053 toises) above today's sea level contain strata of shells and petrified corals. Sometimes, small scattered cones made of basalt, greenstone (*Grünstein*), and porphyric schist crown the tops of high mountains, thus posing difficult enigmas for geologists. The mineralogist can see variations according to the elevation of the ground, just as the naturalist can see variations in the plants and the animals: furthermore, the air, this mixture of gaseous fluids of unknown size enveloping our planet, is

---

3. The greatest vertical depth of mines in Europe is 408 meters (209 toises): the large mine in Valenciana in Mexico has a depth of 516 meters (266 toises).

no less subject to striking variations. As we go further away from sea level, the temperature and pressure of the air diminish; at the same time, its dryness and its electrical tension increase: the blue of the sky seems deeper according to one's altitude. This same altitude also influences the decrease of gravity, the temperature of boiling water, the intensity of the sun's rays traversing the atmosphere, and the refraction of the rays as they travel through it. Thus the observer who leaves the center of the earth by an infinitely small amount compared to the radius can reach a new world, so to speak, and he can observe more variations in the aspect of the soil and more modifications in the atmosphere than he would if he were to pass from one latitude to another.

These variations are found in every region where nature made mountain ranges and high plateaus above sea level; but they are less prominent in temperate zones than at the equator where the Cordilleras have an altitude of 5,000 to 6,000 meters (2,565 to 3,078 toises), and where there is a uniform and constant temperature at each elevation. Near the north pole there are mountains almost as colossal as those found in the Quito kingdom, and whose grouping has all too often been attributed to the earth's rotation. Mount Saint Elias, situated on the American coast opposite the Asian coast, at 60°21' of boreal latitude, is 5,512 meters high (2,829 toises); Mount Fairweather, situated at the 59th degree of boreal latitude, is 4,547 meters high (2,334 toises).[4] In our average latitude of 45 degrees, the Mont-Blanc has a height of 4,754 meters (2,440 toises), and one can consider it to be the highest mountain in the Old Continent, until brave explorers can measure the range of mountains in northwest China, which some have affirmed to be higher than Chimborazo. But in the northern regions, in the temperate zones at 45 degrees, the limit of permanent snow, which is also the limit for all organized life, is only at 2,533 meters above sea level (1,300 toises). The result is that on mountains in temperate zones, nature can develop the variety of organized beings and meteorological phenomena on only half the surface offered by tropical regions, where vegetation ceases to exist only at 4,793 meters (2,460 toises). In our northern latitudes, the slant of the sun's rays and the unequal length of the days raise the temperature in the mountain air so much that the difference between the temperature in the plains and the temperature at 1,500 meters is often imperceptible: for this reason,

4. *Viaje al estrecho de Fuca* [Voyage to the Straight of Fuca], by Don Dionisio Galeano y Don Cajetano Valdes; p. lxv.

many plants that grow at the foot of our Alps are also found at great heights. The rigors of the cold temperature during autumn nights does not destroy their organization; they would undergo the same decrease in temperature a few months later in the plains. A few alpine plants in the Pyrenees grow at very low elevations in the valleys; there they find a temperature to which they might be exposed sometimes also at a higher elevation.

In the tropics, on the contrary, on the vast surface of up to 4,800 meters, on this steep surface climbing from the ocean level to the perpetual snows, various climates follow one another and are superimposed, so to speak. At each elevation the air temperature varies only slightly; the pressure of the atmosphere, the hygroscopic state of the air, its electrical charge, all these follow unalterable laws that are all the more easy to recognize because the phenomena are less complicated there. As a result, each elevation has its own specific conditions, and therefore produces differently according to these circumstances, so that in the Andes of Quito in a region with a breadth of 2,000 meters (1,000 toises) one can discover a greater variety of life forms than in an equal zone on the slopes of the Pyrenees.

I have attempted to gather in one single tableau the sum of the physical phenomena present in equinoctial regions, from the sea level of the South Sea to the very highest peak of the Andes. This tableau contains:

The vegetation;
The animals;
Geological phenomena;
Cultivation;
The air temperature;
The limit of perpetual snow;
The chemical composition of the atmosphere;
Its electrical tension;
Its barometric pressure;
The decrease in gravity;
The intensity of the azure color of the sky;
The weakening of light as it passes through the strata of the
    atmosphere;
The horizontal refractions, and the temperature of boiling water at
    various altitudes.

In order to facilitate the comparison of these phenomena with those of temperate zones, we have added a great number of elevations measured at various locations on the planet, along with the distance at which these elevations can be seen from the sea, not taking into account the earth's refraction.

This tableau contains almost the entirety of the research I carried out during my expedition in the tropics. It is the result of a large number of works that I am preparing for the public and in which I will develop what I can only outline here. I dared to think that this essay would be interesting not only for physicists but even more for all those interested in general physics, to whom it may suggest further comparisons and analogies. This science, no doubt one of the highest achievements of human knowledge, can progress only by individual studies and by connecting together all the phenomena and productions on the surface of the earth. In this great chain of causes and effects, no single fact can be considered in isolation. The general equilibrium obtaining in the midst of these disturbances and apparent disorder is the result of an infinite number of mechanical forces and chemical attractions which balance each other; and while each series of facts must be examined separately in order to recognize a specific law, the study of nature, which is the main problem of general physics, demands the gathering together of all the knowledge dealing with modifications of matter.

I thought that if my tableau were capable of suggesting unexpected analogies to those who will study its details, it would also be capable of speaking to the imagination and providing the pleasure that comes from contemplating a beneficial as well as majestic nature. So many objects are capable of seizing our imagination and lifting us to the most sublime considerations: the multitude of forms developed on the slopes of any one of the Cordilleras; the variety of living structures adapted to the climate of each elevation and to its barometric pressure; the layer of perpetual snow that poses insurmountable obstacles to the spreading of vegetation, but whose limit at the very equator is at 2,300 meters (1,200 toises) higher than under our climates; the volcanic fire that sometimes makes its way to the surface in low hills like Mount Vesuvius, sometimes at elevations five times higher, like the cone of Cotopaxi; the petrified shells found on the peaks of the highest mountains recalling the great catastrophes of our planet; finally, those elevated regions of the atmosphere where the aeronautical physicist went,[5] guided by his bold courage and his noble zeal. By speaking both to our imagination and our spirit at the same time, a physical tableau of the equatorial regions could not only be of interest to those in the field of physical sciences, but could also

5. Mr. Gay-Lussac.

stimulate people to study it who do not yet know all the pleasures associated with developing our intelligence.

In stating these ideas, I have been concerned not so much with the tableau that I am presenting here, the imperfections of which I am well aware, but more with the breadth of which this kind of work is capable. The public, who is so well disposed towards me, will be indulgent toward this work which has been written in the midst of many very heterogeneous occupations. If the new projects I am preparing leave me enough time, I hope to be able eventually to bring this tableau to a higher degree of perfection: for botanical maps are like the ones we call exclusively geographical; we can make them more exact only inasmuch as we accumulate a greater number of good observations.

I sketched out this tableau for the first time in the port of Guayaquil, in February 1803, when I was returning from Lima by way of the South Sea, and was preparing myself for the Acapulco navigation. I sent a first draft of this sketch to Mr. Mutis, who is well disposed toward me, in Santa Fé de Bogotá. No one was more qualified than he to ascertain the accuracy of my observations, and to take them further by means of those he conducted himself in the course of his travels throughout the kingdom of New Grenada for over forty years. Mr. Mutis, a great botanist who followed the results of our science despite his being far away from Europe, observed tropical plants at every elevation. He botanized in the plains of Cartagena, on the banks of the Magdalena River, and in the hills of Turbaco that are adorned with *Gustavia augusta, Anacardium caracoli*, and *Nectandra sanguinea*. He lived for a long time on the high plateaus of Pamplona, of Mariquita, and Ibagué, whose serene skies and delicious climate will always bring back to me the most pleasant of memories. He climbed the snowy peaks of the Andes, those icy regions where grow *Escallonia myrtilloides, Wintera granatensis*, and *Befaria*, which is constantly laden with flowers and which one could call the rose of the Alps of these regions. Mr. Mutis, whose barometric measurements enabled him to ascertain the elevation of these stations, was better able than any other botanist to gather together interesting observations on the geography of plants. Mr. Haenke, who accompanied the unfortunate Malaspina in his navigations, must have made a large number of observations similar to mine: this indefatigable botanist has been living for the past ten years in the high mountain chain of the Andes of Cochabamba that links the Potosí mountains to those in Brazil. Messrs Sessé and Mocinô, who brought to Europe the abundant plant life of Mexico, will no doubt have observed the great variety of plants growing in the soil of New Spain, from the coast of Yucatán and Vera Cruz to the snowy peaks of Citlaltépetl (Orizaba peak) and Popocatépetl. But my stay in the United States and in Mexico as well as a few other circumstances prevented me from benefiting from the advice of these distinguished scientists whose insights could have been very useful to me.

The sketch that I drew in Guayaquil was executed in Paris by Mr. Schoenberger, whose rare talent is known in France and in Germany and who has bestowed on me for many years a special friendship. Having little free time, he was not able to execute it in all the detail necessary for engraving, and so Mr. Turpin was kind enough to take on the task of doing this tableau that I am now offering to the public. Equally distinguished as a painter and as a botanist, he executed this geography of plants with the good taste characterizing all his work. A drawing that by nature is bound to respect scales cannot be done in a very picturesque fashion: all the demands of geodetic precision are contrary to this. Vegetation should be seen as a mass, similar to that depicted in a military map. However, I thought that in the regions closest to the sea, one could represent a grove of scitaminales and palm trees with their slim trunks rising upwards. In this tableau, the eye can see the limits of these regions: there are fewer and fewer palm trees among the other trees, and the trees are gradually replaced by herbaceous plants, and these are displaced by grasses and cryptogams. Some persons of taste might have wished not to see observations surrounding the picture of the Cordillera and to have all these observations relegated near the scales in the margins of the tableau; but in a work of this kind, one must consider two conflicting interests, appearance and exactitude. The public will judge whether we have succeeded in any way in overcoming the difficulties hindering the execution of this sketch.

The tableau of the equatorial regions contains all the physical phenomena occurring on the surface of the earth and in the atmosphere from the 10th degree of boreal latitude to the 10th degree of austral latitude. Extending this zone closer to the borders of the tropical regions would have resulted in less exactitude, because of the great difference one observes not only in the productions of the soil but especially in meteorological phenomena, between the 10th and 23rd degrees of latitude.[6]

6. It will be useful to note here that throughout this work, unless otherwise indicated, the centigrade thermometer scale has been used, and, for linear measurement, the meter, but for time measurements and degrees of latitude, the older terms have been used.

According to the geodetic measurements I executed in Mexico, the limit of perpetual snow goes down only as far as 4,600 meters (2,400 toises) at the 19th degree of boreal latitude, that is, 200 meters (100 toises) lower than at the equator. But many factors give to the regions situated between 20 and 23 degrees of boreal latitude a climate and a type of vegetation that one could not expect to find in the tropics: the proximity of temperate zones, the currents maintaining themselves in the atmosphere, the direction of the trade winds according to the hemisphere where they blow, and many other reasons due to the configuration of the continents. In New Spain the pine trees found at an altitude of as much as 3,934 meters (2,019 toises), and 1,000 meters (500 toises) below the limit of perpetual snow, can still have trunks one meter thick; while under the 5th and 6th degrees of latitude, the taller trees stop growing already at 3,508 meters of altitude (1,800 toises). On the island of Cuba, the temperature sometimes goes down to zero in winter, often for several days. At sea level, it stays only at 7 degrees centigrade, while in Vera Cruz and Santo Domingo, in somewhat more austral latitudes, it does not go below 17 degrees. In the kingdom of New Spain, snow has been known to fall in the capital of Mexico, and even in the Michoacán province, in Valladolid, although these cities' altitude is only 2,264 meters (1,163 toises), and 1,870 meters (959 toises) above sea level. From the equator up to the 4th degree of latitude, it snows only above 4,000 meters (2,000 toises).

Given these data regarding the vegetation and the climate of the regions near the temperate zones, it would not be prudent to show in the same tableau the phenomena occurring in the entire tropical region. Beyond the 10th degree of latitude, north or south, the ground and the atmosphere no longer have the same character as in the equatorial regions.

In my picture, these regions are represented by a vertical cross-section that traverses the high Cordillera of the Andes from east to west. On one side, on the west, one can see the level of the South Sea, which in these regions deserves the name of Pacific Ocean; for from the 12th degree of austral latitude to the 5th degree of boreal latitude, and only within this range, its tranquility is never disturbed by impetuous winds. From the western coast to the Cordillera a long plateau stretches out that is very long from north to south, but only 20 to 30 leagues wide from west to east: this is the Peruvian valley that, north of 4°50′ of boreal latitude, has a vegetation as rich as it is majestic, but is arid and devoid of plants south of this parallel. Covered with granitic sand, sea shells, and rock salt, the soil shows all the characteristics of a land that was submerged by ocean waters for a long time. In this valley, from the hills of Amotape to Coquimbo, the inhabitants have never known either rain or thunder, while north of these hills it rains abundantly and thunderstorms are as furious as frequent. I placed the cross-section of the Andes Cordillera through the highest peak, situated at 1°27' of austral latitude and 0°19' west of the city of Quito: this is the peak of Chimborazo, measured only approximatively by French scientists. Mr. de La Condamine, whose travel narrative contains the most beautiful descriptions of geology and general physics, says that Chimborazo is about 6,274 meters high (3,220 toises); the Spanish geodesist Don Jorge Juan found it to be 6,586 meters high (3,380 toises): this is a considerable difference, since it amounts to 312 meters (160 toises). According to the beautiful map of the Peruvian coasts published by Depósito Hydrográfico in Madrid, Malaspina's expedition judged Chimborazo's height to be 7,496 varas (6,352 meters or 3,258 toises). A geodetic measurement that I did near the new city of Riobamba in the large volcanic plain of Tapia, gives to Chimborazo, supposing a refraction of one-fourteenth of the arc, an altitude of 3,640 meters (1,868 toises) above the Tapia plateau; however, Mr. Gouilly, using Mr. Laplace's formula to calculate the barometric observations, found that this plateau is 2,896 meters (1,485 toises) above sea level: consequently, the total height of Chimborazo would be 6,536 meters (3,354 toises). Using the new formula for refraction that Mr. Laplace was kind enough to furnish me and that he will publish very soon, I found the result of my geodetic measurements changing to 3,648 meters (1,872 toises), and the total height of Chimborazo to be 6,544 meters (3,358 toises). This figure is closer to Don Jorge Juan's estimate than to Mr. de La Condamine's; but one must remember that the latter geodesist perhaps used Bouguer's barometric formula[7] and did not correct for temperature, and so had to find the height to be 180 meters (92 toises) less than my measurements for which I used these corrections. Thus the difference in the assumptions relative to the height of the barometer above sea level leads us further away from being able to mea-

7. The great differences found between the measurements of the same mountains carried out by French and Spanish scientists, differences that are greater than the ones that might result from the uncertainty regarding the Caraburu signal, leads one to suppose that the estimates of Chimborazo's height was modified by different hypotheses about barometric calculation. If, however, the absolute height of all peaks depends on the geodetic measure of the Ilinissa pyramid carried out from Niguas, as the *Figure de la Terre* [Description of the Earth] by Bouguer indicates, then these differences will be even less surprising. I will discuss elsewhere the sources of the errors occurring in this complicated operation.

sure absolute height. Measurements carried out in the Andes Cordillera can be only half geodetic and half barometric, and this complication prevents a comparison between two sets of measurements carried out by very different methods. The length of my base, 1,702 meters (873 toises), the care taken to level it out, and the nature of my angles, should give some confidence in the result of my measurement. Chimborazo's summit is a large segment of a circle, a dome that somewhat resembles the Mont-Blanc. It was impossible to show its shape properly on the plate accompanying this volume; but I am preparing a picturesque image of this colossal mountain whose contours I measured with a sextant, and which I will publish some day.

In this tableau, behind Chimborazo rises a cone 5,752 meters high (2,952 toises); this is the Cotopaxi summit, one of the most active volcanoes in the Quito province along with Tungurahua and Sangay. It is almost five times the height of Mount Vesuvius, which is only 1,197 meters high (615 toises); but it is not the highest volcano on the planet; higher still is Antisana, 5,832 meters high (2,993 toises), with several small mouths, from one of which I saw smoke rise in 1802. In reality, Cotopaxi is not as close to Chimborazo as it seems in my picture. If one had wanted to preserve the true horizontal distances, and if, as in the *Geographical Atlas* that I will soon publish, one wanted to represent the irregularity of the ground in a given region, one would have had to show the Cariguairazo volcano instead of Cotopaxi, for Cariguairazo is a mountain that leans against Chimborazo after having collapsed on July 19, 1698. Cariguairazo is not very interesting today, showing only the ruins of its former grandeur; but I had a very powerful reason for preferring Cotopaxi. I heard the underground groans of this volcano when I was in the port of Guayaquil undertaking the first sketch of this tableau. The mouth of Cotopaxi[8] was at a distance of 42 nautical leagues, yet its explosions sounded like repeated artillery gunfire. In 1744, this volcano was heard as far away as Honda and Monpós, cities 220 leagues hence. If Vesuvius had had the same volcanic intensity, one would have been able to hear it as far away as Dijon or Prague. The height reached by the smoke above Cotopaxi is not arbitrary; it is congruent with the measurements done by Mr. de La Condamine, who estimated that in 1738 the flames reached an altitude of over 900 meters (461 toises) above the mountain peak. During these same

8. Cotopaxi's crater is almost 930 meters wide (478 toises); Rucupichincha's has a diameter of about 1,463 meters (751 toises), while Mount Vesuvius's crater is only 606 meters wide (312 toises).

explosions, this volcano, like others in the kingdom of Quito, vomited immense quantities of fresh hydro-sulfuric water, carbonated clay with sulfur, and fish, barely disfigured by the heat and which comprised a new species[9] of the genus *Pimelodus.*

It is almost superfluous to add that the projection of the Cordillera is represented to scale only for heights, but that this same scale cannot be used for distances. The highest mountains are so small compared to distances that, for example, Chimborazo would be only 4 millimeters (2 lines) high on an in-folio picture that should represent a terrain 200 leagues long; a mountain as high as Mount Vesuvius would even become completely invisible. Furthermore, in order to use the same scale for distance as the one I have chosen for height, and to represent not the whole of South America, but just the small valley contained between the South Sea and the eastern slope of the Andes, one would need a sheet of paper 40 times as long as the format of this book; consequently when one represents a large part of the planet in profile, the scales for height and distance cannot be identical: this factor makes it impossible to show the nature of the terrain because it makes the slopes appear much steeper than they are in reality. I will soon have the opportunity of discussing the advantages and disadvantages of these projections, either in my *Essai sur la pasigraphie minéralogique* [Essay on the System of Mineralogy] or in the *Atlas géologique* [Geological Atlas] that I intend to publish as soon as my astronomical and geodetic observations are sufficiently verified.

The eastern slope of the Andes appears in this tableau a bit less steep than the western slope: thus nature has built this part of the Cordillera through which the cross-section was made. I am quite unconvinced that this general shape is as widespread as Buffon and other famous physicists have believed. When one considers how little known is the eastern slope of the Andes, and how easy it is to confuse the lateral ranges with the high ridges separating the immense plains of the Beni, Purús, and Ucayali Rivers from the narrow Peruvian valley, one must abstain from drawing any general conclusion about the relative incline of the two slopes. When I passed through the Andes Cordillera by the Páramo of Guamani where the Inca had a palace at an elevation of 3,300 meters (1,704 toises) and sketched some constructions similar to Cyclopean ones, when I went down toward the Amazon River and climbed

9. I described *Pimelodus cyclopum* in a separate memoir. See the first notebook of my *Observations de zoologie et d'anatomie comparée* [Observations on Zoology and Comparative Anatomy].

from the province of Jaén of Bracamorros to Micuipampa, I ascertained that at the 3rd and 6th degrees of austral latitude the eastern slope is much less gentle than the one facing the South Sea. Mr. Haenke made the same observation in the Cochabamba province and in the fertile mountains of Chiquitos. Near Santa Fé de Bogotá, the eastern slope of the Cordillera is so steep that no Indian has been able to reach the plains of the Casanare River by way of the Chingasa Páramo.

The crevasse that I showed on the eastern slope of the Cordillera recalls to the observer's imagination one of those narrow valleys that earthquakes seem to have opened in the Andes. Some of them are so deep that Mount Vesuvius, Schneekoppe in Silesia, and the Puy-de-Dôme in Auvergne could be placed there without their peak reaching the height of the mountains nearest to the valley. The Chota valley, in the kingdom of Quito, is 1,566 meters deep (804 toises); the Río Cutacu valley in Peru has a perpendicular depth of over 1,364 meters (700 toises): and yet their floor is still higher above sea level by an equal distance. Their width is often under 1,200 meters (500 toises), and geologists can observe immense seams that nature did not fill with metallic substances. In the Pyrenees also, the Ordesa crevasse near the Mont-Perdu has an average depth of 896 meters (459 toises), according to Mr. Ramond.

At the easternmost extremity of the profile one can see the coasts of the Atlantic ocean, the plains of the Para River, and Brazil. In order to indicate how much longer this part of the picture should be than the rest, a break was made in this immense plain in which flow the Amazon and the Río Negro.

So far I have described the geological phenomena that I have attempted to show in this profile's contours. Let us now cast our eyes toward the interior. Here the geography of the equinoctial vegetation is represented in as great a detail as is possible within the limits of a single plate. Mr. Bonpland and I brought back collections containing over 6,000 species of tropical plants that we gathered ourselves during the course of our botanizing. Since we were also carrying out at the same time astronomical observations as well as geodetic and barometric measurements, our manuscripts contain materials that can determine exactly the position and elevation of these plants. We can show the breadth of the latitudinal zone occupied by these plants, their maximum and minimum elevation, the nature of the soil in which they grow, and the temperature of the plants' native soil.

Following my observations, I placed on the tableau, holding a compass in my hand, the names of the plants that nature put between two specific limits. Each name was entered according to the scale in meters given next to the picture. In order to indicate that the plant is situated on a certain stretch of the Cordillera's slope, the name was often written on a slant. We gave only the generic name when all the known species of the same genus grow at about the same height. Thus, *Escallonia*, *Wintera*, *Befaria*, and *Brathys* are found at the equator only at very great elevations; while *Avicennia*, *Coccoloba*, *Caesalpinia*, and *Bombax* are found only in places near sea level. The restricted framework in which I have given these results allowed me to name only a small number of species: if the public shows some interest in this work, I will follow up with more specialized maps for which all the materials are now ready. How could one show, in one general tableau, 150 species of *Melastoma*, 58 *Psychotria*, 38 *Passifloras*, and over 400 grasses that we brought back from equatorial regions, and which for the most part grow only at specific elevations designated by nature? Often, I was forced to repeat the name of the same genus in order to indicate that some species grow at 500 meters (256 toises) and others at 3,000 meters (1,539 toises). After being back in Europe for a few months, I did not dare to add to this tableau a large number of new genera which we will publish but whose names we are not yet sure about: I have indicated only a few curious plants that are being engraved at this moment and which will be published in the first and second sections of our *Equinoctial Plants*, such as *Cusparia febrifuga* (the valuable tree that gives the *Cortex angosturæ* [angostura bark], a new genus with dull and alternate leaves); *Matisia cordata*, and the wax palm tree (*Ceroxylon andicola*) that Mr. Bonpland described in a separate publication.

In order to bring together the notions one should have about the situation of these plants from a perspective that is more general and more worthy of physics, I have divided this botanical map into regions, according to the analogy existing between forms occurring at different elevations. The names of these regions were engraved in larger letters, as one does for provinces on ordinary maps. Thus, in climbing up from the interior of the earth or from the interior of mines to the icy peaks of the Andes, one can see first *the region of subterranean plants*. These are the cryptogams with an often bizarre structure first studied by Scopoli, and on which I published a separate work (*Florae fribergensis Prodromus, plantas cryptogamicas, praesertim subterraneas, recensens* [Preliminary study of the Freiburg flora, listing the cryptogam plants, especially the subterranean ones] 1790). These are different in specific ways from the cryptogams found on the surface of the earth, and like a great number of them, they seem independent of latitude and climate. Growing in a profound and perpetual darkness, they blanket the walls of subterranean caves and the framework that supports the structures built by miners. I saw the same species (*Boletus ceratophora*, *Lichen verticillatus*, *Boletus botrytes*, *Gymnodermea sinuata*, *Byssus speciosa*) in the mines of Germany, England, and Italy, as in those of those of New Grenada and Mexico, and, in the southern hemisphere, in the mines of Hualgayoc in Peru.

At the same level as these underground cryptogams and in as intense a darkness, some *Fucus* and a few species of *Ulva* also grow, which can be removed with a probe and whose green color presents an interesting problem for physics to account for.

If we leave this multitude of underground plants, we find ourselves in a region where nature took pleasure in bringing together the most majestic forms and grouped them together in a most pleasant sight for the eyes: this is the region where palm trees and scitaminales grow, a region stretching from sea level to an altitude of 1,000 meters (513 toises); this is the homeland of *Musa*, *Heliconia*, *Alpinia*, of the most strongly scented liliaceae, and of palm trees. In this sizzling climate grow *Theophrasta*, *Plumeria*, *Mussaenda*, *Caesalpinia*, *Cecropia peltata*, *Hymenaea*, the balsam of Tolu, and the cusparia or quinquina tree of Carony. On the arid sea coasts, in the shade of coconut trees, of *Laurus persea* and *Mimosa inga*, one can find *Allionia*, *Conocarpus*, *Rhizophora mangle*, *Convolvulus littoralis* and *brasiliensis*, *Talinum*, *Avicennia*, *Cactus*, *Pereskia*, and *Sesuvium portulacastrum*.

Some plants in this region show striking particularities and are remarkable exceptions to the general laws of plant life. Palm trees in southern America, like those of the Old Continent, cannot withstand the cold temperatures of high mountains and cannot grow above an elevation of 1,000 meters (513 toises). One sole Andean palm tree has the extraordinary characteristic of being able to grow only at an elevation equal to that of the Mont-Cenis, and can even be found at altitudes equal to the Canigou. The only palm tree known till now to grow in the Alps, *Ceroxylon andicola*, grows in the Andes of Quindiu and Tolima, at a boreal latitude of 4°25′, at an elevation ranging from 1,860 up to 2,870 meters (954 to 1,472 toises). Its trunk, covered with a wax just analyzed by Mr. Vauquelin, can reach a length of 54 meters.

In the accounts of Admiral Córdoba's expedition, it has been stated that a palm tree was found in the ravines at the Strait of Magellan, i.e. at the 53rd degree of austral latitude. This statement is all the more striking because it is impossible to mistake another plant for the palm tree, unless it is a tree-fern, the existence of which in that Strait would be no less curious. In Europe, *Chamaerops* and the date tree grow only up to 43°40′ of latitude.

Scitaminales, and especially the species of *Heliconia* already described, grow at heights only up to 800 meters (410 toises). Near the Silla of Caracas, at an elevation of 2,150 meters (1,103 toises) above sea level, we found a kind of scitaminales 3 to 4 meters high (9 to 12 feet) and in such abundance that we had great difficulty carving out a path through it: we did not see its blossoms, but according to its overall appearance, it was a new species of *Heliconia* resistant to the low temperatures of these altitudes. *Sesuvium portulacastrum* covers the coasts of Cumaná and grows abundantly in the Perote plain east of the city of Mexico, at an elevation of 2,340 meters (1,200 toises) in a terrain saturated with carbonate and with sodium chloride. Plants in salt marshes seem to me in general to be less sensitive to differences in temperature and barometric pressure.

Above the regions where palm trees and scitaminales are found, are the regions with arborescent ferns and *Cinchona*. The latter is much more widespread than the tree-ferns that grow only in temperate climates, at elevations between 400 and 1,600 meters (205 and 820 toises). Quinquina trees, on the contrary, can grow at an altitude of up to 2,900 meters (1,487 toises) above sea level. The *Cinchona* species that bear up the best in cold temperatures are *Cinchona lancifolia* and Mr. Mutis's *Cinchona cordifolia*: those that grow at the lowest elevations in the plains are *Cinchona oblongifolia* and *Cinchona longiflora*. I found some beautiful trees of the latter species even at an elevation of 740 meters (379 toises). The famous Loja quinquina tree growing in the Caxanuma and Uritucinga forests, very different from the orange-colored quinquina tree of Santa Fé, grows at elevations from 1,900 to 2,500 meters (975 to 1,282 toises). It is a species that presents some analogies with the *Cinchona glandulifera* of the Peruvian flora, but is essentially different. It has so far been discovered only near Loja between the Río Zamora and the Río Cachiyaco, in the Jaén de Bracamorros province, near the village of Sagique, and in a small section of Peru near Huancabamba. It grows on micaceous schists; and in order to set aside completely the inexact name of *Cinchona officinalis*, we will call it henceforth *Cinchona condaminea*, because the illustrious astronomer Mr. de La Condamine first sketched it in that location.

Some travelers stated that they had discovered quinquina trees at elevations of 4,600 meters (2,360 toises), very close to the limit of perpetual snow; but they did not recognize *Wintera* and a few species of *Weinmannia*, whose bark contain abundant tannins and are also used with success as a febrifuge. We saw no true *Cinchona* trees above 2,900 meters nor below 700 meters (1,487 and 359 toises); the quinquina tree in the Philippines described by

Cavanilles, and the one recently discovered on the Island of Cuba in the Guines valley, seem to belong to a different genus.

Rubber is produced by plants that have few similarities, by *Ficus*, *Hevea*, *Lobelia*, *Castilloa*, and several euphorbia. Camphor is also found in plants that do not belong to the same genus: in Asia one finds it in a laurel tree; in Peru, in the fertile Cochabamba province, one can find it in a didynamous bush that Mr. Haenke discovered growing abundantly near Ayopaya. The fruit of *Myrica* and the trunk of a palm tree produce wax. Some substances whose chemical properties are the same are furnished by plants with very different structures: the same applies to the fever-reducing capacities of the quinquina tree that exist in plants not belonging to the same genus.

The cusparia of the Caroní River plains near the city of Upatu, that majestic tree that gives *Cortex angosturæ*, belongs to a very different genus from the *Cinchona*. *Cuspa* or *Cumaná quina*, whose blossoms we have not been able to obtain till now, has alternate leaves without stipules: it does not belong to the genus *Cinchona*, even though it would be difficult for a chemist to tell a *Cuspa* infusion from one of the yellow quinquina of Santa Fé. On the coasts of the South Sea west of Popayán near Atacamez, a tree grows having the properties of *Cinchona* and *Wintera*, and which no doubt is different from both of these genera. The cusparia in Guyana, the *Cuspa* of New Andalusia, and the *Cascarilla* of Atacamez all grow at sea level, and nature prepares in their sap a substance analogous to that of the true quinquina trees growing at an elevation of 2,800 meters (1,436 toises).

In the narrative of my journey to the tropics, I will publish a *botanical map of the Cinchona genus*. It will show the places in both hemispheres where this interesting tree is found. It will show that it grows in the Andes Cordillera in an area over 700 leagues long. One will be able to see the *Cinchona* trees in an area from the Potosí region and the Plata River, situated at the 20th degree of austral latitude, up to the snow-covered mountains of Santa Marta at the 11th degree of boreal latitude. The entire eastern slope of the Andes, south of Huánuco, near the Tipuani mines, in the environs of Apollobamba and Yuracarées, is an uninterrupted forest of quinquina trees. Mr. Haenke saw that it stretches almost up to Santa Cruz of the Sierra. It does not seem that this tree grows any further to the east; it has not been found up till now in the Brazilian mountains, even though the Chiquitos Cordillera seems to link them with the Peruvian Andes. From La Paz on, *Cinchona* continue to grow throughout the provinces of Gualias and Gumalias, and into Huancabamba and Loja. In the east, they grow in the province of Jaén de Bracamorros, and even crown the hilltops near the Amazon River, near the famous Pongo de Manseriche. From Loja on, the quinquina tree goes as far as Cuenca and Alausi in the Quito kingdom: it is abundant east of Chimborazo; but it seems to be missing entirely in the whole high plateau of Riobamba and Quito, as in the Pasto province up to Almaguer. Have the great volcanic catastrophes that frequently wrack this country decreased the number of species? In general, we have observed that vegetation is less varied there than in other regions with the same elevation above sea level. North of Almaguer, which I found to be at 1°51'57" of boreal latitude in the province of Popayán, quinquina trees are again abundant. They grow almost without interruption throughout the Andes of Quindiu, the Vega-de-Supia, the fertile hills of Mariquita, Guaduas, and Pamplona, up to the mountains of Mérida and Santa Marta, where waters of boiling hydro-sulfuric springs mingle with those of melting snows.

Silla-de-Caracas and a few mountains in the Cumaná province (Tumiriquiri, the environs of the Caripé convent, and the Guanaguana mountain pass) have an elevation of 1,300 to 2,500 meters (667 to 1,282 toises), and consequently they provide cool enough temperatures for *Cinchona* to grow there. The same applies to the kingdom of New Spain, whose high plateau has a climate quite similar to Peru's. However, up till now, no one has discovered any *Cinchona* either in the Cumaná province nor in Mexico. Could this phenomenon be accounted for by the scarcity of mountains surrounding the high peaks of the Santa Marta and those of Guamoco? The ridge of the Andes Cordillera disappears almost entirely between the Cupica Gulf and the delta of the Río Atracto. The Panama isthmus is below the lower limit of *Cinchona*. Has this plant, in its migration to the north, encountered obstacles in the too great heat of the climate in those regions? or in time, will not the quinquina tree be found in the beautiful Jalapa forests east of Vera Cruz, where the nature of the soil, the tree-ferns, the melastome trees, the temperate climate, and the humid air all seem to proclaim to the botanist that he will find this beneficial tree he has so far vainly searched for in this region?

In the temperate regions of the *Cinchona*, some Liliaceae grow, for example *Cypura* and *Sisyrinchium*, *Melastoma* with large purple flowers, passiflora trees that are as tall as our northern oaks, *Bocconia frutescens*, *Thibaudia*, *Fuchsia*, and *Alstroemeria* of rare beauty. This is where the *Macrocnemum*, *Lysianthus*, and cucullaires soar majestically. The ground there is covered with *Koelreutera*,

Weissia, Dicranum, Tetraphis, and other mosses that are always green. In the ravines hide Gunnera, Dorstenia, Oxalis, and a multitude of unknown Arum. At an elevation of about 1,700 meters (872 toises) one finds Porlieria hygrometrica, which has been identified thanks to Messrs Ruiz and Pavon, Citrosma with scented leaves and fruit, Eroteum, Hypericum baccatum and cayenense, and many species of Symplocos. Above 2,200 meters (1,129 toises), we have not found any mimosas whose irritable leaves fold upon contact: the cool temperatures of these elevated regions set this limit to their irritability. Above 2,600 meters (1,334 toises) and especially at an altitude of 3,000 meters (1,539 toises), Acaena, Dichondra, Nierembergia, Hydrocotile, Nerteria, and Alchemilla make a thick lawn. This is the region of Weinmannia, oaks, Vallea stipularis, and Spermacoce. There Mutisia climbs on the tallest trees.

In equatorial regions, oaks (Quercus granatensis) grow only at an elevation above 1,700 meters (872 toises). In Mexico, under the 17th and 22nd degrees of latitude, I have seen them grow at altitudes as low as 800 meters (410 toises). Sometimes these oak trees indicate in springtime the changing of the seasons at the equator: they lose all their leaves, and one can see new leaves replacing them, whose young green color mingles with that of Epidendrum growing on their branches.

A new genus of malvaceae, Cheirosthemon, on which Mr. Cervantes, a botany professor in Mexico, published an interesting monograph, is also found in these elevated regions; but its flower, with its bizarre structure, has not been discovered up till now in the Peruvian Andes. For a long time, only one example was known, in the suburbs of the city of Toluca in Mexico. It grows wild in the Guatemala kingdom, and the famous hand-tree of Toluca was most likely planted there by some Rointztèques peoples. The gardens in Iztapalapan, whose remains Hernandez was able to glimpse, are a witness to the taste that the peoples whom we call savage and barbarians had for cultivation and for the beauties of the plant kingdom.

Near the equator, tall trees whose trunk exceeds 20 to 30 meters (10 to 15 toises) do not grow above an elevation of 2,700 meters (1,385 toises). From the level of the city of Quito on, the trees are not as tall and their height is not comparable to that of the same species growing in the most temperate climates. Above 3,500 meters (1,796 toises) there are almost no trees; but at this elevation, bushes are all the more common: this is the region of berberis, of Duranta ellisii and mutisii, and Barnadesia. The plateaus of Pasto and Quito are characterized by such plants, just as the region of Santa Fé is characterized by *Polymnia* and *Datura* trees. *Castelleja integrifolia* and *fissifolia*, *Columella*, the beautiful *Embothryum emarginatum*, and *Clusia* with four anthers are common in this region. The ground is covered with a multitude of calceolaria, whose gold-colored corolla makes a pleasing contrast with the green color of the lawn where they grow. Nature assigned them a particular zone: this zone begins at one degree of boreal latitude. Messrs Ruiz and Pavon, who conducted scientific research in Chile, will be able to indicate how far calceolaria extend in the austral hemisphere. Higher, on the summit of the Cordillera, from 2,800 meters to 3,300 meters (1,436 toises to 1,693 toises), lies the region of *Wintera* and *Escallonia*. The climate of these regions, cold but constantly humid, called *páramos* by the natives, produces bushes whose short, carbon-covered trunks divide into a multitude of branches covered with tough and shiny green leaves. A few orange quinquina trees, *Embothrium*, and *Melastoma* with purple, almost scarlet flowers, can grow at these elevations. *Alstonia*, whose dried leaves makes a beneficial tea, *Wintera granatensis*, and *Escallonia tubar* which stretches out its leaves like an umbrella, grow in scattered groups. Underneath them grow some small *Lobelia*, bassella, and *Swertia quadricornis*.

Still higher, at 3,500 meters (1,796 toises), trees cease to grow, as I stated previously. Only on the Pichincha volcano, in a narrow valley going down from the Guagua Pichincha, have we discovered a group of [singenèses en arbre], with trunks 7 or 8 meters high (21 or 24 feet). From 2,000 meters to 4,100 meters (1,026 to 2,103 toises), this is the region of alpine plants: *Staehelina*, gentians, and *Espeletia frailexon*, whose furry leaves are often used as shelter by unfortunate Indians surprised by night in these regions. The carpet is adorned with *Lobelia nana*, *Sida pichinchensis*, *Ranunculus gusmani*, *Ribes frigidum*, *Gentiana quitensis*, and many other new species of plants that we will describe in our *Plantes équinoxiales* [Equinoctial Plants]. *Molina* are smaller bushes that we encountered at their greatest altitude on the Puracé volcano, near Popayán, and on the Antisana volcano.

At an altitude of 4,100 meters (2,103 toises), alpine plants make way for grasses,[10] whose region stretches up to 4,600 meters (2,360 toises). *Jarava*, *Stipa*, a multitude of new species of *Panicum*, *Agrostis*, *Avena*, and *Dactylis* cover the ground. From afar it looks like a golden carpet, called by the inhabitants of the region *pajonal*. From time to time snow falls on this region of grasses.

10. La Condamine, *Voyage à l'Équateur* [Voyage to the Equator], p. 48.

At an altitude of 4,600 meters (2,360 toises), there are no more phanero-gams at the equator. From this elevation to that of perpetual snow, only li-chenous plants cover the rocks. Some seem to hide even under the perpetual ice; for at 5,554 meters (2,850 toises) of altitude, near the Chimborazo sum-mit, I found on a rocky ridge *Umbilicaria pustulata* and *Verrucaria geographica*: those are the last organized beings we found in the ground at these great heights.

These are the principal characteristics of vegetation in the physical tab-leau of equatorial regions; it would be desirable to have a similar one for Eu-rope. There are so many data not given in the classic works of Messrs Pallas, Jacquin, Wulfen, Lapeyrouse, Schranck, Villars, Host, and a great number of traveling naturalists. The famous botanists who traveled through the Alps of Salzburg, the Tyrol, and Styria, those who saw the high peaks of Switzerland and Savoy, would be able to furnish botanical maps much more complete than the one I am offering today to the public. Who has more precious mate-rials for this work than the one[11] who, on the peaks of the Pyrenees, discov-ered that immense deposit of organic debris, who, equally knowledgeable in geology and botany, possesses both the art of observing well and the talent of speaking to the imagination?

I have already discussed the reasons why the phenomena of the geography of plants cannot be as varied nor as constant at the 45th degree of latitude as they are at the equator. Despite this disadvantage, a *Tableau physique des climats tempérés* [Physical Tableau of Temperate Climates] would nevertheless be very interesting. At its center, one would see the Mont-Blanc in the high Euro-pean mountain range rise to 4,775 meters (2,448 toises). The slopes of this range would stretch on one side toward the Atlantic, and on the other toward the Mediterranean basin, where *Chamaerops*, date trees, and several plants on the Atlas mountains foreshadow the proximity of Africa. In this depiction, perpetual snow would descend to an altitude of 2,550 meters (1,307 toises) above sea level, that is to say, at an altitude where, at the equator, grow wax palm trees, quinquina, and the most vigorous trees. Thus the zone in Europe between the level of the ocean and perpetual snow is half as narrow as in the tropical regions; but the snowcap covering the highest peaks of Europe, the Mont-Blanc and the Mont-Rose, is wider by 600 meters (308 toises) than the

11. The author of *Observations faites dans les Pyrénées* [Observations Made in the Pyr-enees], and of the *Voyages au Mont-Perdu* [Voyages to the Mont-Perdu], Mr. Ramond.

one covering Chimborazo. On the craggy rocks that rise above the limits of perpetual snow and remain bare because of the steepness of their slopes, in the Alps surrounding the Mont-Blanc at more than 3,100 meters (1,590 toises) of altitude, the following plants grow: *Androsace chamaejasma*, Jacq.; *Silene acaulis*, growing as low as 1,500 meters (769 toises) that Saussure found growing at 3,468 meters (1,778 toises); *Saxifraga androsacea, Cardamine alpina, Arabis caerulea*, Jacq., and *Draba hirta* of Villars, which is *Draba stellata*, Wild. At these great heights *Myosotis perennis* also grows, and *Androsace carnea*, whose stems become gradually smaller. The latter finally has only a single blossom and is found above 1,000 meters up to 3,100 meters (513 to 1,590 toises). In the Pyrenees, the highest regions from 2,400 to 3,400 meters (1,231 to 1,744 toises) are adorned with *Cerastium lanatum*, Lam., *Saxifraga groenlandica, Saxifraga androsacea, Aretia alpina*, and *Artemisia rupestris. Cerastium lanatum* does not even grow below 2,600 meters (1,333 toises). In the Alps, from 2,500 meters to 3,100 meters (1,282 to 1,590 toises), on the debris of rocks and gravel surrounding the permanent snow, and on the highest glaciers, the following plants grow: *Saxifraga biflora*, Allion., *Saxifraga oppositifolia, Achillea nana, Achillea atrata, Artemisia glacialis, Gentiana nivalis, Ranunculus alpestris, Ranunculus glacialis*, and *Juncus trifidus*. In the high mountain range of the Pyrenees at 3,000 meters (1,539 toises) and even at 1,500 meters lower (769 toises) are found: *Potentilla lupinoides*, Wild., *Silene acaulis, Sibbaldia procumbens, Carex curvula*, and *Carex nigra*, Allion., *Sempervivum montanum* and *Sempervivum arachnoideum, Arnica scorpioides, Androsace villosa*, and *Androsace carnea*. In the Alps, between 2,300 and 2,500 meters (1,180 and 1,282 toises), at the height reached by the edge of snows and glaciers, these plants grow not on stones but in a fertile soil, in fields moistened by highly oxygenated waters from melting snow, on a lawn of *Agrostis alpina: Saxifraga aspera* and *bryoides, Soldanella alpina, Viola biflora, Primula farinosa, Primula viscosa, Alchemilla pentaphyllea, Salix herbacea* that grows higher than any other woody plant, *Salix reticulata*, and *Salix retusa. Tussilago farfara* and *Statice armeria* also grow from the plains up to 2,600 meters (1,333 toises) of elevation. In the Pyrenees are found at these elevations *Scutellaria alpina, Senecio persicifolius, Ranunculus alpestris, Ranunculus parnassifolius, Galium pyrenaicum*, and *Aretia vitaliana*. Above the lower limit of perpetual snow, between 1,500 and 2,500 meters (769 and 1,028 toises), in the Savoy Alps, grow the following: *Eriophorum scheuchzeri, Eriophorum alpinum, Gentiana purpurea, Gentiana grandiflora, Saxifraga stellaris, Azalea procumbens, Tussilago alpina. Passerina geminiflora, Passerina nivalis, Merendera bulbocodium, Cro-*

cus multifidus, Fritillaria meleagris, and Anthemis montana grow in the Pyrenees at the same elevation. Genista lusitanica, Ranunculus gouani, Narcissus bicolor, Rubus saxatilis, and a number of gentians are found lower. Rhododendrum ferrugineum usually prefers elevations between 1,500 and 2,500 meters (769 to 1,282 toises); however, Mr. Decandolle, to whom I owe these observations on the Alps, also saw it in the Jura mountains, at the bottom of Creux-du-Vent at 970 meters (498 toises) above sea level.

Linnaea borealis, which grows at sea level near Berlin, in Sweden, in the United States, and in the Nootka Sound, grows in the Swiss Alps at 500 and 700 meters (256 and 359 toises) of elevation. One can find it in the Valais along the edges of the torrent flowing under the Tête Noire; on the Saint-Gothard, where Haller saw it first; near Geneva on the Voirons mountain, according to Saussure; and even in France, near Montpellier, on the Espinouse.

Trees whose trunks exceed 5 meters (2.5 toises) grow at the equator at elevations barely up to 3,500 meters (1,796 toises). In the kingdom of New Spain at the 20th degree of latitude, a pine tree similar to Pinus strobus grows as high as 3,934 meters (2,018 toises); oaks grow as high as 3,100 meters (1,590 toises). The naturalist who is not aware of this phenomenon of plant geography would believe, simply from looking at them, that these mountains covered with pines at very high altitudes cannot be equal in height to the Tenerife peak. In the Pyrenees, Mr. Ramond observed that the two trees growing at the highest altitudes on the mountain peaks are Pinus sylvestris and Pinus mugho; they are found at altitudes between 2,000 and 2,400 meters (1,026 and 1,231 toises). Abies taxifolia and Taxus communis begin at 1,400 meters (718 toises) and go as far as 2,000 meters (1,026 toises). Fagus sylvatica occupies an intermediate region, from 600 to 1,800 meters (308 to 923 toises): but Quercus robur, found in the plains, can go only as far as 1,600 meters (821 toises); it stops growing 200 meters (102 toises) higher than the inferior limit for Pinus mugho.

Mr. Ramond[12] also furnished me with very interesting observations on the maximum and minimum altitudes at which species belonging to the same genus are found. Choosing the genera Primula, Ranunculus, Daphne, Erica, Gentiana, and Saxifraga, I present here a table showing the altitudes where each of the species making up these genera grows in the Pyrenees:

12. See also his botanical observations in his Voyage au sommet du Mont-Perdu [Voyage to the Summit of Mont-Perdu], 1803, p. 21; and his Mémoire sur les plantes alpines [Memoir on Alpine Plants] in the Annales d'histoire naturelle [Annals of Natural History].

		Meters	Toises
Gentiana	pneumonanthe	0 to 800	0 to 400
	verna	600 – 3000	300 – 1540
	acaulis	1000 – 3000	500 – 1540
	campestris	1000 – 2400	500 – 1200
	ciliata	1200 – 1800	600 – 900
	lutea	1200 – 1600	600 – 800
	punctata, Villars	1600 – 2000	800 – 1000
Daphne	laureola	300 – 2000	150 – 1000
	mezereum	1000 – 2000	500 – 1000
	cneorum	2000 – 2400	1000 – 1200
Primula	eliator	0 – 2200	0 – 1100
	integrifolia	1500 – 2000	750 – 1000
	villosa	1800 – 2400	900 – 1200
Ranunculus	aquatilis	0 – 2100	0 – 1050
	gouani	500 – 2000	250 – 1000
	thora	1400 – 2000	700 – 1000
		1500 – 2400	750 – 1200
	alpestris	1800 – 2600	900 – 1300
	amplexicaulis	1800 – 2400	900 – 1200
	nivalis	2000 – 2800	1000 – 1400
	parnassifolius	2400 – 2800	1200 – 1400
	glacialis	2400 – 3200	1200 – 1640
Saxifraga	tridactylides	0 – 40	0 – 20
	geum	400 – 1600	200 – 800
	longifolia	800 – 2400	400 – 1200
	aizoon	800 – 2400	400 – 1200
	pyramidalis	1200 – 2000	600 – 1000
	exarata	1400 – 1800	700 – 900
	cespitosa	1600 – 3000	800 – 1540
	oppositifolia	1600 – 3400	800 – 1740
	umbrosa	1400 – 1800	700 – 900
	granulata	1200 – 1600	600 – 800
	groenlandica	2400 – 3400	1200 – 1740
	androsacea	2400 – 3400	1200 – 1740
Erica	vagans	0 – 900	0 – 450
	vulgaris	0 – 2000	0 – 1000
	tetralix	500 – 2400	250 – 1200
	arborea	550 – 700	270 – 350

The saxifrages in Tyrol present the same characteristics as those of the Pyrenees. Count Sternberg, who botanized in these mountains and on the Baldo and to whom we owe a description of its geology, communicated to me an interesting note on the Rhododendrums and other alpine plants. I owe it to these botanists and to the physicists to insert this note here in its entirety. Mr. Sternberg writes:

Unless there is a particular local circumstance, the region where *Rhododendrum* grows is hardly ever below 876 to 974 meters (450 to 500 toises). I did not find them lower than 100 meters (50 toises) above the Wallersee in Bavaria, which is 817 meters (420 toises) above sea level. *Rhododendrum chamaecistus* does go as low as *ferrugineum* and *hirsutum*. Besides, I have found them growing equally on primary limestone and on secondary limestone, in the Sette *communi* and on Mount Sumano that is 1,277 meters high (656 toises): they accompanied me up to an altitude of 1,950 meters (1,000 toises).

The most extensive region of alpine saxifrages appears to me to be in the Tyrolean Alps. I found *Saxifraga cotyledon* and *aizoon* in the Eiszach valley between Brixen and Botzen at 360 meters (184 toises) of altitude. They accompanied me till the Grappa summit, near Bassano, at 1,684 meters (865 toises). *Saxifraga caesia*, *aspera*, and *androsacea* are found in the intermediate region; above, one finds *Saxifraga autumnalis*, *mucosa*, *moschata*, and *petraea*; the highest-growing ones are usually *Saxifraga burseriana* and *bryoides* that cover the Baldo summit at 2,225 meters (1,143 toises). Primulas, especially *farinosa*, *auricula*, *marginata*, and *viscosa*, are not found in the Tyrolean Alps below 801 meters (417 toises). Representing a singular anomaly, *Primula farinacea* grows in the Regensburg plain. As for *Ranunculus glacialis* and *Ranunculus seguierii*, I never observed them growing below 1,950 meters (1,000 toises) of elevation.

However, in order to complete the geography of plants, it would not suffice to compose tableaux of the regions near the poles, the ones with temperate climates from 40 to 50 degrees of latitude, and the equatorial ones; it would not suffice to describe the austral and the boreal hemispheres, because the plants in Chiloé and Buenos Aires are very different from the ones in Spain and Greece: one would also have to furnish separate tableaux for the New and the Old Continents. Madagascar with its high granitic peaks perpetually covered in snow according to Commerson, with its coasts so well researched by Mr. du Petit-Thouars, the Adam peak on Ceylon, the island of Sumatra, where the Ophyr cone rises, according to Marsden, to an altitude of 3,949 meters (2,027 toises), all these could furnish precious materials for the tableaux of the equatorial regions of Africa and the East Indies. The illustrious Pallas might be able to determine the geography of plants in the temperate climates in Asia. Mr. Barton, who is equally versed in zoology, botany, and the study of Indian languages, is currently conducting this type of research in the temperate zones of the United States. There the mountains are no higher than 2,000 meters (1,026 toises);[13] the height of 3,100 meters (1,582 toises) attributed to the White Mountain in New Hampshire by Messrs Cutler and Belknap is no doubt exaggerated. Mr. Barton does not find in his homeland the same variety of phenomena present in the highest of the Cordilleras; but this lack is largely compensated by the great variety of arborescent plants found in the beautiful plains of Pennsylvania, Carolina, and Virginia. In the United States, there are almost three times the number of oak species as the number of different species of large trees present in all of Europe. The aspect of vegetation in the New Continent is more varied and more pleasant than in the old at the same latitude. Gleditschia, tulip trees, and magnolias contrast in a most picturesque manner with the dark foliage of Thuia and the pines: one can imagine that nature took pleasure in adorning a soil that would be inhabited one day by an energetic and industrious people worthy of enjoying peacefully all the benefits given to them by a free society.

But the physical tableau of the equinoctial regions is useful not only for developing new ideas regarding the geography of plants; I believe that it could also help us understand the totality of our knowledge about everything that varies with the altitudes rising above sea level. This consideration encouraged me to bring together in 14 scales many numbers resulting from the large quantity of research conducted in various branches of general physics. These scales being self-explanatory, it will be sufficient to add only few words on their composition. The scales indicating the temperature, the air's hygroscopic state and its electrical tension, the blue color of the sky, the geological aspects, the cultivation of the soil, and the diversity of animals according to altitude, all these are based on my observations made during my expedition, which will be developed in further detail in the account of my travel to the equator.

13. See the work of Mr. Volney containing a broad description of the structure of the planet in the northern part of the New Continent.

THE TEMPERATURE SCALE

This scale shows the maximum and minimum temperatures that the centigrade thermometer indicates in intervals of 500 meters (250 toises). These data are the result of many thousands of observations carried out over five years, often hour by hour. The average temperature is not the middle temperature between these extremes, but the middle temperature among all the observations made at such and such an altitude. One has tried to not confuse the effects of a general law with what seems to depend only on different locations. For example, in this table, one can see that at sea level, the temperature does not go below 18.5°, although in Havana it has been observed several times to go down to +1.4° and even to zero; but this city is 13 degrees further away from the equator than the zone whose phenomena I am describing here; and when the north winds blow violently, the proximity of the continent results in cold temperatures not usually found at this latitude. On the island of Santo Domingo, a little further south, the thermometer hovers constantly between 23 and 24 degrees in the plains. It is superfluous to note that all the observations of temperatures were made in the shade and far from the reflection of ambient heat.

Heights above sea level meters	toises	Maximum temperatures		Minimum temperatures		Average temperature	
0 to 1000	0 to 500	+	38.4°	+	18.5°	+	25.3°
1000 to 2000	500 to 1000	+	30.0	+	12.5	+	21.2
2000 to 3000	1000 to 1500	+	23.7	+	1.2	+	18.7
3000 to 4000	1500 to 2000	+	20.0	±	0.0	+	9.0
4000 to 5000	2000 to 2500	+	18.7	−	7.5	+	3.7
5000 to 6000	2500 to 3000	+	16.0*	−	10.0*	−	0.2*

The numbers indicated in this scale for altitudes above 5,000 meters (2,500 toises) are not very precise: this elevated region has not yet been explored very much, and for too few hours, to ascertain completely its average temperature. The cold temperature indicated by the thermometer on the peaks of the Andes is never very considerable, although this temperature is difficult to endure because of the diminished quantity of oxygen, the depression of the nervous system, and other so far unknown reasons. In their cabin in Pichincha at an altitude of 4,735 meters (2,428 toises), the scientists saw the centigrade thermometer go down only to 6° below zero, and on Chimborazo at an altitude of 5,908 meters (3,032 toises), my thermometer showed only −1.8°. On the large Antisana volcano at the great altitude of 5,403 meters (2,773 toises), even in the shade the thermometer reached 19°. On the contrary, in the places known to be the hottest on earth, Cumaná, Guyara, Cartagena in the Indies, Guayaquil, situated on the coasts of the South Sea, along the Magdalena River and on the banks of the Amazon, the average temperature is 27°, while in Paris and Milan it ranges from 11° to 13°. But in these same equatorial regions the thermometer rarely reaches the extremes of heat that one so often sees in northern Europe. Upon examining a table of over 21,000 observations carried out with very good instruments by Mr. Orta, a marine officer in the service of the King of Spain, I found that for over 13 years in Vera Cruz the centigrade thermometer went only three times[14] over 32° and never over 35.6°. On the contrary, in Paris, the temperature often reaches 36°, and on April 14, 1773 it was observed to reach 38.7°. In Vera Cruz the average temperature for the months of May, June, July, August, and September is 27.5°, and I found that the cruel adynamia fever known as *vómito prieto* ravages the region every time the temperature goes over 23.7°. In equatorial regions, the difference between the extreme high and low temperatures is 16 to 20 degrees. In Europe, at the 5th degree of latitude, the difference is over 62 degrees on the centigrade thermometer.

Because the ground heats up a great deal on the seacoasts as in the immense plains of the Orinoco, herbaceous plants with very short stems, *Sesuvium*, *Gomphrena*, *Thalinum*, *Killinga*, and a few mimosas half buried in sand, can tolerate a temperature of 52 degrees. In the Jorullo plains in Mexico, I saw plants growing in a black sand that made the thermometer go as high as 60 degrees during the day. On the contrary, *Staehelina*, *Swertia*, and other plants growing on the Andes peaks endure a constant temperature of +3.5°, with the exception of a few hours per day that they are warmed by the sun.

14. Mr. Wilson (*History of the British Expedition to Egypt*, p. 134) states that in Egypt on May 22 1802, the centigrade thermometer reached 53 degrees in the shade at Belbeis during the sirocco. If this is exact, the sand present in the atmosphere must have contributed to raising the air temperature.

These alpine plants and palm trees occupy, so to speak, the two extremes of this botanical thermometer.

The average temperatures given in a scale from 1,000 to 2,000 meters (500 to 1,000 toises) indicate the decrease in heat at the equator from sea level to the peaks of the Andes. If the selection of the observations on which I based these average temperatures were correct, the decrease in heat that would result would be more exact than any observations that could be made in Europe above 3,000 meters (1,500 toises), the latter observations being very few in number and very isolated. Voyages undertaken to the summits of the Alps, or ascents in balloons, can never be frequent enough to ascertain the exact average temperature of the air strata from 3,000 to 5,000 meters (1,500 to 2,500 toises). In the tropics, on the contrary, there are villages at elevations of more than 400 meters (200 toises) above the Tenerife peak, where a physicist can find it not too painful to stay and study meteorology profitably.

From my observations in the Andes Cordillera, I saw that the decrease in heat is, in a ratio of 5:3, quicker above 3,500 meters (1,750 toises) than from sea level to 2,500 meters (1,250 toises). The stratum of air where the decrease in temperature at the equator is the quickest seems to be between 2,500 and 3,500 meters (1,250 and 1,750 toises), or between the heights of Mount Saint-Gothard and Etna. It is easy to understand how radiant heat, being modified by the irregularities of the earth's surface or by the shape of the mountains, can influence this decrease. A physicist making a balloon ascent at the equator above the plains of the Amazon might find perhaps very different temperatures from the ones I observed on the slopes of the Cordillera; but it is probable that this difference would not go much beyond 4,000 meters (2,000 toises), an altitude where, in these Andes, the mass of the mountains and consequently their influence on the ambient air are considerably diminished.

The voyage that I made towards the Chimborazo summit indicated a decrease of heat of 196 meters (98 toises) for one degree on the centigrade thermometer. The average temperatures on the scale show that it is 189 meters (100 toises), going from sea level to an altitude of 5,500 meters (2,823 toises). Saussure believes that in Europe the decrease in heat was 156 meters (90 toises) for one centigrade degree in the summer, and 233 meters (111 toises) in the winter. In his great balloon ascent, Mr. Gay-Lussac observed a decrease in heat in the summer identical to the one I observed at the equator. This scientist observed a temperature of zero at 5,000 meters (2,500 toises) (the temperature in Paris being 30 degrees), while at 6,000 meters (3,000 toises) it was 3 degrees below freezing. These data situate the decrease to be 183 meters (92 toises) from 0 to 5,500 meters. But in calculating the entire column of air traversed by Mr. Gay-Lussac, one finds, from 0 to 6,977 meters, the decrease to be 173 meters (87 toises) for each centigrade degree. In my *Memoir* on the inferior limits of perpetual snow, I showed that above 4,700 meters (2,300 toises) of elevation, the difference in latitude seems to have very little influence on the temperature, and that, during his last climb, Mr. Gay-Lussac encountered above this level, at 48 degrees of latitude, air strata having exactly the same temperature as the strata where I was at the same height on Chimborazo. The phenomena of horizontal refraction, 4 or 5 minutes smaller at the equator than in Europe, seem contrary to the temperatures being equal in elevated regions. They indicate a decrease in heat that is quicker at the equator than what my observations state; but it must be noted that, according to Mr. Delambre, horizontal refractions in Europe are not as strong as generally believed. The phenomenon of refractions being dependent on the totality of the air strata traversed by the rays, an uneven decrease in regions above 7,000 meters (3,500 toises), never explored by anyone, can cause differences in horizontal refraction that Bouguer observed at the equator. Thus, our uncertainty about the decrease in heat during European winters, and the lack of congruence[15] between le Gentil's and Bouguer's observations, prevent us from obtaining reliable data, and I must be content for the moment with stating the facts as I observed them in equatorial regions.

THE BAROMETRIC SCALE

This scale shows the air pressure in the atmosphere at different elevations above sea level, as expressed by the level in the thermometer. These elevations were calculated according to the barometric formula that Mr. de Laplace published in his *Mécanique céleste* [Celestial Mechanics], based on the average

15. Indeed, Mr. Delambre believes that there is only a very slight difference between horizontal refractions in temperate zones and in the tropics. He carefully recalculated all the observations made by le Gentil in Pondichéry, in which Borda had found a mistake in reduction. These calculations proved to Mr. Delambre that the refractions are the same in Europe as in the Indies. However, le Gentil's observations seem very exact.

temperatures that are shown by the temperature scale. Let X represent the height in meters, H the level of the barometer at sea level, T the temperature at the same level, t the temperature corresponding to the height X, and h the height of the barometer sought for the elevation X; one has:

$$Log(m) = x/18393\{1 + 2(T + t)/1000\}$$

and having found the number m, one has:

$$h = H/m\{(1 + T - t)/5412\}.$$

This formula results in the following barometric levels at 500-meter intervals:

Elevations above sea level			Barometric Levels	
Meters m	Toises t	Temperatures in centigrade	Meters m	Lines (based on the Paris foot) l
0	0	+ 25.3°	0.76202	337.8
500	256	+ 24.0	0.71961	319.03
1000	513	+ 22.6	0.67923	301.18
1500	769	+ 21.2	0.64134	284.28
2000	1026	+ 20.0	0.60501	268.24
2500	1282	+ 18.7	0.57073	253.05
3000	1539	+ 14.4	0.53689	238.06
3500	1795	+ 9.0	0.50418	223.50
4000	2052	+ 6.4	0.47417	210.20
4500	2308	+ 3.7	0.44553	197.55
5000	2565	+ 0.4	0.41823	185.40
5500	2821	− 3.0	0.39206	173.84
6000	3078	(− 6.0)	0.36747	162.95
6500	3334	(− 10.0)	0.34357	152.38
7000	3591	(− 13.0)	0.32035	142.61
7500	3847	(− 16.0)	0.30068	133.36

The average temperatures above 6,000 meters (3,000 toises) are not quite exact: they are based only on the hypothetical law of the decrease in heat.

Mr. de Saussure observed the barometer going down to 0.43515 meters (16 inches 0.9 lines) on the summit of the Mont-Blanc. On the Corazón peak (south of the city of Quito) Messrs de La Condamine[16] and Bouguer observed a barometer registering 0.42670 meters (15 inches 9.2 lines). I carried instruments on Chimborazo at such an altitude that I saw the mercury go down to 0.37717 meters (13 inches 11.2 lines); but in his balloon ascent Mr. Gay-Lussac endured a dilation of the air corresponding to 0.3288 meters (12 inches 1.7 lines).

The barometric level at sea level was determined to be only 0.76202 meters (337.8 lines), the temperature being 25 degrees centigrade. This is what the observations I made in the tropics indicated, both on the Atlantic and the South Sea coasts. Bouguer said it was 0.76022 meters (28 inches 1 line), and the Spanish geodesist, Don George Juan, 27 inches 11.5 lines. La Condamine says that "if the average level of the barometer is not less than 28 inches in the tropics, the difference is very slight." My observations, made with barometers well cleansed of air by means of fire, and compared to those in the Paris observatory, seem to prove that the average pressure of the air at sea level in the tropics is a little less than in temperate zones.

Mr. Schuckburg[17] found the latter to be 0.76300 meters (28 inches 2.24 lines): Mr. Fleuriau Bellevue, 0.76434 meters (28 inches 2 5/6 lines) at a temperature of 12 degrees. This difference of almost 2 millimeters cannot be explained solely by the difference between average temperatures in Europe and in equatorial regions; even less so because in the low parts of Peru, when the sun is hidden for four to five months behind a thick fog, the centigrade thermometer remains between 15 and 16 degrees. This is as insoluble a problem as the hourly barometric oscillations at the equator, which I no longer dare to consider to be like tides in the aerial ocean, since I ascertained that the moon has almost no influence on them.

The elasticity of the air in temperate zones varies in the same location sometimes by as much as 0.0450 meters (20 lines). In the tropics, where the

16. Mr. de La Condamine states: "No one ever saw the barometer go so low in the open air, and it is unlikely that anyone ever climbed to a greater altitude; we were at 4,815 meters (2,471 toises), and we vouch for the exactitude of this measurement, give or take 8 or 10 meters (4 or 5 toises)" (*Voyage à l'Équateur* [Voyage to the Equator], p. 58).

17. It would be important to determine very exactly the average level on the coasts of the Mediterranean and the Atlantic Ocean.

trade winds constantly carry air strata with even temperatures from the 10th degree north to the 10th degree south of the equator, this elasticity does not vary at the edge of the sea by more than 0.0026 meters (1.4 lines), and at 3,000 meters (1,500 toises) of elevation it does not vary by more than 0.0015 meters (0.7 lines). But even though the amount of these variations is very small, they are nevertheless remarkable by the law followed by the barometer's movement from hour to hour. Godin was the first to point out this phenomenon, without noting the periods of the maximum and minimum barometric levels. Mr. La Condamine noted these periods to be between 9 o'clock in the morning and 3 o'clock in the afternoon. Mr. Balfour in Calcutta and Mr. Moscley in the Antilles also noted the time periods; but they do not correspond to the ones I found with Mr. Bonpland, when we stayed up for several nights in a row to observe the nocturnal tides. We found that the barometer was at the highest level at 9 o'clock in the morning, that it fell very little till noon, but a lot more between noon and 4 or half past 4 in the afternoon; that it rose again till 11 o'clock at night when it was a little lower than at 9 o'clock in the morning. It fell again during the entire night till 4 o'clock in the morning, when it was a little higher than at 4 o'clock in the afternoon. Then it rose from 4 to 9 o'clock in the morning. The time periods of these hourly variations are the same on the coasts of the South Sea and in the Amazon plains, and in the elevated regions at 4,000 meters (2,000 toises). They seem to be independent of changes in temperatures and seasons. If the mercury is trending downward from 9 o'clock to 4 o'clock in the afternoon, and if it is trending upward from 4 o'clock to 11 o'clock at night, neither a storm, nor an earthquake, nor the strongest rains and winds can alter its course. Nothing seems to determine its course except real time and the position of the sun. In some places in the tropics, the moment when the mercury begins to fall is so noticeable that it indicates real time within one quarter of an hour. At sea level at the equator, the average level of the barometer being $= z$, I can determine its level to be approximately:

at 21 h. $= z + 0.5$ at 11 h. $= z + 0.1$

at 4 h. $= z - 0.4$ at 16 h. $= z - 0.2$

Among the many thousands of observations that we made of the barometer's hourly oscillations, I cite only one example that can illustrate this type of regularity. The direction of the arrows indicates whether the barometer is rising or falling.

Observations made in the Port of Callao, near Lima, on November 8 and 9, 1802: the barometer[18] is equipped with a vernier with which one can easily discern 0.03 lines.

Hours real time		Barometer (in lines)	Thermometer fixed to the barometer	Thermometer in the open air in the shade	Deluc's Hygrometer
	hours				
November 8 at	10 ½	336.92	19	16.3°	43°
	11	336.98	19	16.2	43.7
	13	336.72	19.5	16.2	44
	14	336.60	19.5	16.2	42
	15	336.65	19.8	16.5	43
↘	15 ½	336.62	20.0	16.0	42
	16	336.55	19.0	16.0	42
	16 ½	336.80	20.5	16.3	42.5
↗	17	336.87	22.0	16.4	42
	17 ½	336.95	22.7	17.0	42
	20	337.25	23.0	18.0	39
	21	337.35	23.0	19.2	37
	22 ½	337.13	24.5	20.4	37.5
	hours				
November 9 at	0 ½	336.90	25.5	22.5°	34°
	0 ¾	336.75	25.9	22.7	34
↘	3 ½	336.60	26.0	23.0	34.5
	4	336.45	25.5	20.5	33.6
	5	336.50	25.5	18.0	37
	8	336.85	25.0	16.1	39
	9	336.95	22.0	16.5	40
↗	10	336.97	22.4	16.4	42
	11	337.15	20.0	16.4	42
	11 ½	336.90	20.5	16.7	42
	13	336.84	20.5	17.0	43

18. This observation was made 12 meters (6 toises) above the level of the South Sea. The level of the barometer not having been rectified precisely, the absolute heights are about 0.9 lines too small.

Mr. Mutis, who spent over 30 years observing these hourly oscillations, believed that he observed, at an elevation of 2,623 meters (1,347 toises) in Santa Fé de Bogotá, that the conjunctions and the oppositions of the moon influenced the barometric tides. I could not observe these changes; however, I do not doubt their existence. Mr. Laplace calculated the effect of the influence of the sun and the moon on the aerial ocean; but perhaps it is masked at the equator by the phenomenon of hourly variations. In the boreal hemisphere near the tropics, the cold north winds, which blow very hard in the Gulf of Mexico, make the barometer rise by 5 to 7 lines. This extraordinary phenomenon, one of the most important indicators for navigation between Havana and Vera Cruz, is an entirely local one between the 19th and the 23rd degrees of latitude. This mass of cold air makes the mercury rise, interrupting the pattern of hourly oscillations; but they begin again at Vera Cruz as soon as the storm is over. Mr. Cotte made a large number of observations in Europe, from which he deduced that the minimum level of the barometer occurs 2 hours after the sun's reaching its highest point, and therefore 2 hours earlier than at the equator. In our temperate climates, the hourly variations of the weight of air are hidden perhaps by a multitude of local causes that make the barometer go up and down irregularly: but I do not doubt, agreeing with Mr. Van Swinden, that averages deduced from thousands of observations made from hour to hour indicate that, even in our latitudes, the barometer rises and falls at specific periods.

I cannot end these discussions about air pressure without adding a physiological observation. In the city of Quito, the barometer remains at 0.54366 meters (20 inches 1 line). In the city of Micuipampa, I found it to be 0.49629 meters (18 inches 4 lines), and the inhabitants of the Antisana district breathe an air whose elasticity is expressed in a column of mercury of 0.46927 meters (17 inches 4 lines). Mr. Gay-Lussac observed dips in the barometer down to 0.3288 meters (12 inches 1 line). The person who is used to a pressure of 0.76 in the plains (28 inches) can tolerate all these changes. The inhabitants of these cities situated at high elevations enjoy a very good health, even though newcomers have at first a bit of trouble breathing, especially if they are talking fast or exerting some strong muscular effort, small discomforts that last only a very short time. This malaise becomes more pronounced, however, when the barometer falls below 0.40605 meters (15 inches). At elevations of 5,000 meters (2,500 toises), the nervous system feels very weakened. One faints easily with the least effort. Sometimes people feel like vomiting, and above 5,800 meters (2,900 toises) of elevation, moving the muscles and the lack of air pressure often act upon blood vessels having very thin walls, so that one bleeds from the eyes, lips, and gums. These phenomena vary with the travelers' constitution; some people even do not experience them at all. Saussure observed that man is more resistant to a rarification of air than mules. I led a horse to the Perote Cofre to an altitude of 3,839 meters (1,970 toises). Its breathing was very belabored. It also seemed to me that the white race suffers less above 5,800 meters (2,900 toises) than the race of indigenous people with copper-colored skins.

Air pressure must have the greatest influence on the vital functions of plants and especially on the breathing of their teguments. Although many cryptogams, and among the phanerogams especially the grasses, are indifferent to these modifications in barometric pressure, others however are not. Swertia quadricornis, Espeletia frailexon, Chuquiraga, and a few gentians seem to require a dilation of air equal to 0.46 or 0.49 meters (17 or 18 inches). If many plants were transplanted from the Andes to the equally cold regions of Europe, they would not grow to their fullest because they would not find that rarified air to which their organs are accustomed in their native location. The considerable changes observed in the physiognomy of plants when transplanted to the plains are attributed solely to differences in temperature, humidity, and electrical tension. But I do not see why barometric pressure would be excluded, which is probably just as powerful an influence on the organization of plants. In living nature, many causes contribute to the modification of vital functions, and none can be neglected to explain the phenomena of the organization of matter.

THE HYGROMETRIC SCALE

This scale shows the decrease in the humidity of the atmosphere as one rises above sea level. The observations used to determine these averages were made in the shade under a clear blue sky. Sometimes I used Saussure's hygrometer, sometimes Deluc's, according to whether the instrument was to absorb humidity quickly or whether it could be left exposed for a long time to the air which it was to measure. All the results were reduced to the scale of Saussure's hygrometer, correcting them for temperature, and reducing them to 25.3° of the centigrade thermometer. Saussure's and Dalton's experiments prove that there is no need to correct the barometer.

Elevations	Saussure's Hygrometer, not corrected for temperature	Thermometer attached to the hygrometer	Hygrometer reduced to the temperature of 25.°3
meters			
0 to 1000	86	+ 25.3	86
1000 to 2000	80	+ 21.2	73.4
2000 to 3000	74	+ 18.7	64.5
3000 to 4000	65	+ 9.0	46.5
4000 to 5000	54	+ 3.7	36.2
5000 to 6000	38	+ 3.0	26.7

These averages shed some light on the decrease in humidity in equatorial regions, a decrease which is not without relevance for research on refractions. This decrease is 90 meters (45 toises) per degree on Saussure's hygrometer.

Despite the extreme dryness of the air on the peaks of the Andes, where the hygrometer goes down to 46 degrees with a thermometer at 3.7° (which amounts to 31.7° on the hygrometer, the thermometer being at 25.3°), in these regions at 2,500 to 3,500 meters of elevation (1,250 to 1,750 toises) one is constantly enveloped in a thick fog. These precipitations of water, which are either the effect or the cause of a strong electrical tension, give to the vegetation of the *páramos* a characteristic freshness of color.

In the low tropical regions, the air is completely transparent and without the trace of a cloud for 4 to 5 months of the year, and it is considerably laden with water. Mr. Deluc proved, using his son's experiments, that this great humidity also exists in Bengal. This hygroscopic state of the air sustains the plants in times of drought. If these plants did not have the capacity to pull vaporized water from the air, how could they conceivably sustain themselves in places such as Cumaná, where for 8 to 10 months there is no rain, no mist, and no dew?

At an elevation of 2,350 meters (1,175 toises) in the Mexican valley, Saussure's hygrometer often goes down to 42 to 43 degrees, with a thermometer showing 15° to 18°. In Europe, I have never observed a dryness above 46 degrees, with a temperature of 15 degrees. But what in the

Mexican valley absorbs the vapors rising from the five lakes surrounding the capital? This absorption cannot be explained by the immense quantity of sodium chloride and sodium carbonate covering the ground. The whole interior of the kingdom of New Spain is astonishingly dry. Vegetation is very rare above 2,000 meters (1,000 toises) of elevation, and the air seems artificially dried, so to speak. This dryness, no doubt as harmful to health as to vegetation, increases from century to century, because human manufacturing drains the lakes and the rains are no longer abundant. How dry the air must be in Persia, between Tiflis and Tabriz and in the Kerman province, where, according to Chardin, houses are built out of rock salt!

Water vaporized in the air, and precipitated out of it because of a change in temperature, or perhaps for other insufficiently known causes, forms groups of air cells that we see as clouds. Their elevation, which I have often measured, seems rather constant. The lower layer of these clouds seems to me to be 1,169 meters (600 toises) above sea level. At this elevation on the slope of the Cordillera, there is during part of the year a constant, thick, enveloping fog in Jalapa, east of Mexico, and in Guaduas, in the kingdom of Santa Fé. The upper limit of the large clouds is about 3,300 meters (1,600 to 1,800 toises); but a very interesting phenomenon is the existence of small clouds commonly called moutons [woolly], at over 7,800 meters (3,900 toises) of elevation. We saw them above us on the Antisana volcano, and Mr. Gay-Lussac also mentions them in his account of his second balloon voyage. How light are these air cells, capable of maintaining themselves in such a rarified atmosphere! According to Messrs Biot's and Gay-Lussac's observations, the lower limit of clouds in Europe during the summer seems to be 1,200 meters (600 toises), as at the equator.

At an elevation of 5,267 meters (2,635 toises), Mr. Gay-Lussac saw the hygrometer at 25.5°, with the thermometer being at +4°. This is no doubt the maximum dryness that has ever been observed: because, after reducing the hygrometer to the temperature of 25.3° that exists in the plains during the summer, the 25.5° are reduced to 21.5°.

The amount of rain falling in the tropics is over 1.89 meters (70 inches) per year. In Guayaquil, in the Cumanacoa valley, and between the Casiquiare and the Río Negro, I think I can estimate it to be 2.43 meters (90 inches). In the United States, at the 40th degree of latitude, the quantity of rain is 1.80 meters (40 inches); in Europe, 0.48 meters (18 inches).

THE ELECTROMETRIC SCALE

When one goes from sea level to the summit of the Cordilleras, one sees the electrical tension increasing gradually, while on the contrary heat and air humidity are observed to decrease more and more. The experiments given in this table were carried out at different times of the day by means of Saussure's electrometer equipped with a conductor 1.4 meters high (4 feet), transmitting electricity in the atmosphere by means of smoke from punk, as was proposed by Mr. Volta. In the low-lying equatorial regions from sea level up to 2,000 meters (1,000 toises), the lower strata of air are not charged with much electricity. One can scarcely find any signs of it after 10 o'clock in the morning even with Bennet's electrometer. All the fluid seems to accumulate in the clouds, which causes frequent electrical discharges, usually occurring at regular intervals two hours after the sun reaches its highest point, when the heat is at its maximum and the barometric tides are near their minimum. In the valleys of great rivers, for example in the valleys of the Magdalena, the Río Negro, and the Casiquiare, storms occur constantly around midnight. In the Andes, electrical discharges are the strongest and the noisiest between 1,800 and 2,000 meters (900 and 1,000 toises): the Caloto and Popayán valleys are known for a frightful frequency of these phenomena. Above 2,000 meters (1,000 toises), storms are less frequent and less regular; but they contain a lot of hail, especially at 3,000 meters (1,500 toises) of elevation, because the air there is frequently and for a long time negatively charged, which is almost never the case, or at the most only for a few fleeting instants, below 1,000 meters (500 toises). Above an elevation of 3,500 meters (1,750 toises), discharges are quite rare; there is hail without lightning, and very often, above 3,900 meters (1,950 toises), it falls mixed with snow and even in the middle of the night. The air strata near the high peaks of the Andes have a constant electrical tension manifested on Saussure's electrometer by 4 to 5 lines. The activity of electricity is easier to feel because of the dryness of the air and the proximity of the clouds. Near the mouths of volcanoes, electricity often changes from positive to negative. In the area above perpetual snow a great number of light events occur that do not seem to be accompanied by thunder. The multitude of shooting stars seen coming down in the volcanic part of the Andes, and their greater frequency in warm climates, might lead one to consider that these events belong to our planet, if other reasons, especially their great elevation, were not contrary to this idea.

THE AZURE COLOR OF THE SKY

When the inhabitants in the plains climb to heights of 3,000 to 4,000 meters (1,500 to 2,000 toises), they are struck by the dark hue of the heavens' azure vault. The intensity of the color increases with the dilation of the atmosphere and the decrease in the vapor masses traversed by the sun's rays. The dispersion of light produced by the vesicular vapor makes the sky greyish or milky. The lesser the air mass traversed by the sun's rays, the darker the sky, which becomes almost as black as it would appear to us if we were situated at the upper limits of the atmosphere.

The cyanometer that I used in this expedition was made by Mr. Paul in Geneva, modeled on the one used by Saussure on the Mont-Blanc. These observations were made at the zenith.

It appeared to me that in general the blue of the sky is more intense in the tropics than at an equal height in Europe. The average for Paris (the summer temperature being 25°) seemed to me to be 16° on the cyanometer. In the tropics, I estimated it to be 23°. This difference seems to be caused probably by the perfect dispersion of vapors in the equatorial atmosphere. Thus nothing comes close to the majesty of the nights in these regions: fixed stars shine with a steady light quite similar to that of the planets; any scintillation is seen only very close to the horizon. Some weak optical instruments carried from Europe to the Indies seem stronger there, so great and constant is the transparency of the air.

On the summit of the Mont-Blanc at 4,754 meters of elevation (2,438 toises), Saussure observed the cyanometer registering 39°. On the Tenerife peak I saw it at 41°. The great dryness of the African air increases the sky's color; for Tenerife is 1,500 meters (540 toises) lower than the Mont-Blanc. In the Andes, at 5,900 meters (3,000 toises), the cyanometer registered 46°. Mr. Gay-Lussac observed this same intensity of color in his balloon expeditions.

This physicist said to me:

A phenomenon that struck me at the great height of 7,016 meters (3,508 toises) was the existence of clouds above me, and at a distance which seemed indeed very considerable. In our first ascent there were no clouds above 1,169 meters (600 toises), and above me the sky was extremely pure. Its color at the zenith was so intense that it could even be compared to Prussian blue. But in the last voyage I just concluded, I did not see any clouds below me; the sky was very vaporous, and its color generally lackluster.

THE DECREASE IN LIGHT

The light emitted by the sun and the stars is weakened as it passes through the atmosphere. This decrease in light depends on the density of the air strata; consequently it is weaker on the summits of high mountains and stronger at sea level. In calculating the following table, vapors occurring accidentally in the air were not taken into account. The phenomenon of the light's decrease to extinction was considered as it might occur in a transparent atmosphere in which water is perfectly vaporized. On this topic, one can consider the ideas stated by Mr. Laplace in his *Système du monde* [The System of the World] (vol. 1, p. 157). Because of the great transparency of the air in the tropics, the light there is either brighter or less weakened than in Europe even at an equal height. How tiring the great clarity of the day in the Indies seems to be, even when there is no reflection? It would also be interesting to examine this phenomenon with Leslie's photometer. The lessening of the decrease of light in the tropics is also manifested very strikingly by the light that the moon emits during a total lunar eclipse and directs toward the earth; this light is due to the inflection of the sun's rays passing through the terrestrial atmosphere. In temperate zones, the air is often so dense and so filled with vapors that the lunar disk disappears completely. But at the 10th degree of boreal latitude, the atmosphere is so transparent that I have seen the light of the eclipsed moon appear almost as bright as does the full moon in our regions when it starts rising above the horizon.

It is well known that light exerts a powerful influence on the vital functions of plants, especially on their breathing, on the formation of their coloring part that has a resinous character, and, according to Berthollet,[19] on the fixation of nitrogen in their starch. These considerations lead us to suspect with reason that the great intensity of light to which plants are exposed on mountain peaks must contribute to the resinous and aromatic nature of a large number of alpine plants. In my work on nerves, I cited some experiments in which solar light seems to produce stimulating effects on nervous fibers that would be difficult to attribute solely to heat. The feeling of weakness felt by the inhabitants of Quito and Mexico, whenever the sun strikes them hard at 3,000 to 4,000 meters of elevation (1,500 to 2,000 toises), does not seem to be correlated with muscular exertion or perspiration through the skin, which no doubt increases in the dilated air. Would this feeling be due to an irritation of the nerves? Or does light, less weakened on

---

19. *Statique Chimique* [Chemical Statics], vol. 2, p. 496.

mountain summits, give off more heat as it is broken down by dense bodies, because it has lost even less in passing through?

HORIZONTAL REFRACTIONS

Since the refractive force of the atmosphere depends on the density of its layers and on the law of their temperature, this force is different according to the elevation where the observer is located. Mr. Laplace proved that the numbers for astronomical refractions are very different depending on whether the angle being observed is above or under 12 degrees. In the first case, the hygroscopic state of the air hardly modifies the inflection of light; in the second case, when the rays strike the earth almost horizontally, the influence of the watery vapors and their more or less even dispersion becomes more important. If the decrease in heat alone modified horizontal refractions, it would be difficult to conceive why they are by far much smaller at the equator than in temperate zones during the summer; for, according to the experiments quoted above, it is likely that in the summer the decrease in heat, at least from the surface of the sea up to 6,000 or 7,000 meters of elevation (3,000 to 3,500 toises), is not very different in the Andes near Quito than in Europe. But perhaps the Cordilleras, as they reflect radiating heat into the higher air layers, do not give enough comparable results, or perhaps the decrease is different above 7,000 meters of elevation. It is extremely important to observe these phenomena exactly, as they are of great interest for physical astronomy, phenomena on which Mr. Laplace's new work will shed much light. This great geodesist's formulas are the basis for my scale of refractions that adorn my *Physical Tableau of Equinoctial Regions*.

On a marble tablet still in front of the former Jesuit college in Quito, French scientists had the average astronomic refraction engraved as being 27′ at sea level and at the equator; at the elevation of Quito, 22′50″; and on Chimborazo near the lower limit of perpetual snow, 19′51″. Mr. Laplace states that the thinness of the moon's atmosphere being far superior than any vacuum that our best pneumatic machines can produce, the horizontal refraction on the moon's surface cannot go above 5 seconds.

On the high peaks of the Andes, sometimes one sees in the middle of the night a pale but distinct light surrounding the horizon. Saussure observed it on the Col-de-Géant at 3,435 meters of elevation (1,717 toises). I glimpsed it sometimes, especially on the Antisana farm, at 4,105 meters (2,523 toises). Mr. Biot offered an ingenious explanation for this phenomenon, which he attributed to the reflection of solar light caused by the thick and deep mass of air of the edge of the horizon (*Astronomie physique* [Physical Astronomy], vol. 1, p. 277).

THE CHEMICAL COMPOSITION OF THE ATMOSPHERE

The elastic fluid enveloping our planet reaches heights whose limits we do not know. The theory of the extinction of light, as well as Bouguer's experiments, prove that the height of the atmosphere, reduced in all its extension to the density of air at a temperature of zero and a pressure of a column of 0.76 meters (28 inches) of mercury, would be 7,820 meters (3,910 toises) (*Mécanique céleste* [Celestial Mechanics], vol. 4). Observations carried out at dusk show that at 60,000 meters (30,000 toises), the density of air strata is still enough to reflect some perceptible light.

For a long time it has been thought that not only the chemical composition of the atmosphere was variable in the same place but that the purity of the air diminished as one rose above sea level. What was attributed to the modifications of the air was only the result of using imperfect eudiometric tools. The experiments I conducted with nitrous gas contributed in no small way to spread these errors.

In the past few years, it has been proclaimed that the quantity of oxygen contained in air, far from being 27 or 28 hundredths, was only 21 to 23 hundredths. These limits are still too imprecise, and since chemists are still uncertain as to the quality of their eudiometric tools, Mr. Gay-Lussac and I have undertaken an extensive investigation of the composition of air and the modifications that it can undergo. I wanted to replace an imperfect study done in my youth with one based on more solid foundations.

What is true for chemistry is also true for astronomy. The perfection of our methods and instruments allows us to determine the smallest quantities, and today one can no longer ignore what seemed unmeasurable yesterday. Mr. Gay-Lussac and I published our first results in a memorandum read at the Institute on the 1st of Pluviôse Year XIII [January 21, 1805]. The eudiometric numbers indicated on my table are based on the experiments we did in one of the laboratories of the École Polytechnique, which we hope to follow up in a more varied and extensive manner.

Given the present state of our knowledge of chemistry, Volta's eudiometer is preferable to other eudiometric tools. It is the only one that can register changes in the air of 2/1000 of oxygen. Alkaline sulfur, phosphorus, and nitrous gas (obtained by washing the residues with iron sulfate and oxygenated hydrochloric acid and alkali) can evaluate the quantity of oxygen with a precision of only one or two hundredths. Alkaline sulfur, when heated, absorbs nitrogen, and by attributing all the absorption to the oxygen in the atmosphere, sulfur often seems to indicate that there is 30 to 40 hundredths oxygen. Because of this action of sulfur dissolved at a high temperature, and because of erroneous suppositions about the saturation of one part of oxygen to 2 to 4 parts of nitrous gas, statements were made in the past that there was 0.27 to 0.28 oxygen in the air.

The atmosphere appears to be constituted of 0.210 oxygen, 0.787 nitrogen, and 0.003 carbonic acid gas. The quantity of the latter has not been ascertained with the necessary precision. It might be even smaller. The alkaline solutions generally used probably do not act only upon carbonic acid; because every time a liquid remains in prolonged contact with air, the absorption of nitrogen and oxygen can modify the results.

The chemical composition of the atmosphere does not seem to vary with respect to its proportion of oxygen and nitrogen. If variations do exist, it is unlikely that they go beyond 1/1000 of oxygen; whether the air was collected during a rainy period, in fog, in dry and quiet conditions, or during a period of snow and wind coming from all directions, it always contained 0.210 or 0.211 oxygen. Mr. Gay-Lussac determined the important fact that at 7,000 meters (3,500 toises) of elevation, the atmosphere also contains 0.21 oxygen. That is the only experiment made with high precision on the chemical composition of the highest strata of air. If other explorers and I believed that there was less oxygen at higher elevations than at sea level, one must suspect that the imperfection of our eudiometric tools misled us. On the summit of Tenerife and on a few volcanoes in the Andes, the air might indeed be less pure; but this difference must no doubt be attributed to the action of their craters, and especially to the great masses of sulfur that absorb oxygen from the air with which they are in immediate contact.

The question, an important one, whether air contains hydrogen has been debated. During his second balloon ascent, Mr. Gay-Lussac proved that if there is a small quantity of hydrogen in the air, it is not greater at 7,000 meters (3,500 toises) of elevation than in the plains. We have just performed further research on this topic, and we can state that there cannot be more than 2/1000 hydrogen in the atmosphere; for our instruments measured 0.003 dispersed in an artificial mixture of oxygen and nitrogen. Now, a mixture of air containing less than 0.05 hydrogen does not catch fire with an electrical impulse, and so it seems that hydrogen contained in the atmosphere cannot explain the formation of rainstorms and other phenomena of ignition. The constant uniformity of the chemical composition of the air and the lack of hydrogen are two very important facts for calculating refractions. They prove that geodesists do not need to any other corrections than those of the barometer, the thermometer, and the hygrometer.

But aside from oxygen and nitrogen, the atmosphere contains a large number of gaseous emissions that our current instruments do not indicate and that can impact powerfully on our health. These emissions are produced particularly in the low-lying regions of the tropics, where organic matter is produced more rapidly, and where the same organic debris fills the air with putrid and deleterious miasmas. The humidity of the air and its constantly elevated temperature and the absence of wind in the shade of the forests are favorable for the formation of these miasmas. They are particularly abundant in the yawning valleys of the Andes that resemble crevasses with depths of 1,200 to 1,500 meters (600 to 750 toises), where radiant heat registers 42 degrees on the thermometer. Staying there for one hour can suffice to cause very serious illnesses for travelers, while the Indians inhabiting these valleys are used to these miasmas and enjoy a perfect and constant good health. Such is the admirable organization of man.

THE DECREASE IN GRAVITY

Gravity decreases as one moves away from the center of the earth. This decrease can be felt even on the small elevations of the Cordilleras; but since these mountains have a very different density, I prefer to determine the decrease in gravity through theory rather than through the experiments that I conducted in all too divergent circumstances. This scale shows the oscillations of a simple pendulum in a vacuum.

If the length of the seconds pendulum observed in Paris is = 1.000000, at the equator it will be = 0.99669. These relationships depend on the dimensions of the earth: the radius of the equator = 6,375,703 meters (3,271,208 toises); the radius at the pole = 6,356,671 meters (3,261,443 toises); flattening = 19,032 meters (9,765 toises); the length of one degree (at the equator) = 51,077.70 toises, Bouguer; in France, at a latitude of 51°332 = 51,316.58 toises, Méchain and Delambre; in Sweden, at a latitude of 73°707 = 51,473.01 toises, Melanderhielm.

Let N be the number of oscillations of a pendulum placed at the equator and on the surface of the earth; let N' be the number of oscillations of the same pendulum carried vertically to the height h: this height being expressed in meters, one has:

$$N' = N\{1 - 579h/576.6375793\}$$

One might be surprised that my tableau does not mention the decrease in magnetic force at very great elevations. But the elegant experiments of Messrs Biot and Gay-Lussac have proven sufficiently well that this decrease is not perceptible from sea level to 6,000 meters (3,000 toises) of elevation. The observations carried out on the summit of the Cordilleras are affected by local attractions. By making the needle of my inclination meter oscillate on the mountain of Guadeloupe at 676 meters (338 toises) above the plains of Santa Fé, I observed 2 fewer oscillations, in 2 minutes' duration, than in the plain. On the Cerro of Avila, near Caracas, at 2,632 meters (1,316 toises) above sea level, the decrease was as much as 5 oscillations; and on the contrary on the Antisana volcano at an altitude of 4,934 meters (2,417 toises), the number of oscillations during 10 minutes was 230, while in the city of Quito, it was only 218: this indicates an increase in intensity. These anomalies can result only from local conditions. One can read on this topic the paper that I have just published with Mr. Biot on the variations in the earth's magnetism.

THE TEMPERATURE OF BOILING WATER AT VARIOUS ALTITUDES

The amount of heat accumulated by liquids before boiling depends on the weight of the atmosphere, and since this weight varies with the elevations above sea level, each elevation has its corresponding boiling point. The following table explains this phenomenon.

		Boiling temperature of water	
Elevation in meters	Barometric altitude	Centigrade thermometer	Réaumur thermometer
0^{meters}	0.7620°	100.0°	80.0°
1000	0.6792	97.1	77.7
2000	0.6050	94.3	75.4
3000	0.5368	91.3	73.0
4000	0.4741	88.1	70.5
5000	0.4182	84.7	67.7
6000	0.3674	81.0	64.8
7000	0.3203	77.0	61.6

During my voyage I did a large number of experiments on the temperature of boiling water on the summit of the Andes. I will publish some others done by Mr. Caldas, a native of Popayán, a distinguished physicist who with incomparable ardor studied astronomy and several branches of natural history. These experiments, not very interesting theoretically, can be useful only for determining the exactitude of altitude measurements done with a thermometer, if one were to possess instruments indicating small fractions of a degree with precision. From sea level to 7,000 meters (3,500 toises), one degree in the decrease of the temperature of boiling water occurs every 304 meters (152 toises); but from zero to 1,000 meters, one degree is the equivalent of 357 meters (185 toises). One can suppose that up to the altitude of the Mont-Blanc, one degree of decrease in the temperature of boiling water indicates approximately 10 lines of barometric decrease, or 340 meters of elevation.

GEOLOGICAL CONSIDERATIONS

The nature of rocks is generally independent of differences in latitudes and altitudes, either because the air temperature and its barometric pressure had little influence on the agglomeration of molecules, or because the formation of the earth's solid sphere occurred before the order of things existed that assigned a climate particular to each region. Thus the height of the tallest mountains is so insignificant in relation to the radius of the earth that these small differences in level have had no effect on vast geological phenomena. When one considers the earth's sphere in general, one could almost believe that any rock can be found at any elevation.

But the influence of altitude can be seen when one considers a small part of the earth's surface. Then one discovers that in each region the direction and the slant of each layer were determined by a particular set of forces,[20] and that there is a local law for the altitude to which the various rock formations rise above sea level. Thus in a given region, one can see that secondary

20. In the Andes of South America, in the Venezuelan Cordillera, and in the one in Pavia, primary rocks, especially laminated granite and micaceous schist, often take the direction 3 4/8 hora on the miner's compass, that is, the layers lie at an angle of 52 degrees from north to east, along the local meridian. Their slant is almost always toward the northwest. This direction and this slant of schistous rocks are also very common in the Swiss Alps, in the Fichtelgebirge, and on the coasts of Genoa. In Mexico, the primary rocks' direction is hora 7–8 on the Saxony compass.

mountains do not rise above 3,000 meters (1,500 toises); that the limestone masses are not covered with sandstone above 1,800 meters (900 toises); that micaceous schist is not found as high as laminated granite, and that every outcropping that rises above a certain height is composed only of primary rocks. On a given small terrain, for example, one can find an upper limit of the basalts, of secondary limestone, or sandstone with a silicon base, just as one finds upper limits for pines or oaks. From these considerations, it follows that one cannot make a geological scale for the equatorial regions, unless one wishes to model nature according to theoretical ideas, in other words, to consider as general some phenomena occurring only in a very small part of the Andes; however, I thought it would be interesting for the mineralogist that my tableau contain some geological considerations.

The equatorial regions in America have both very high peaks and very extended plains that are the lowest in the world, forming a contrast that shows that the earth's rotation did not cause this cluster of mountains near the equator. Thus, at the 60th degree of boreal latitude, the Andes Cordillera rises again to an altitude almost equal to that observed in the kingdom of Quito.

The mountain chain of the Andes, whose Peruvian name is *antis*, deriving from *anta*, meaning copper, stretches almost equally toward each of the earth's poles. Its extremities are only 29 to 30 degrees of latitude removed from the poles. It stretches from the small islands south of the Tierra del Fuego, or Cape Horn, up to Mount Saint Elias, situated north-west of the port of Mulgrave, that is, from the 55°58′ of austral latitude up to 60°12′ of boreal latitude. It is 2,500 leagues long and 30 to 40 leagues wide.

The elevation of the Andes Cordillera is much more unequal than commonly believed. There are parts of it in the austral hemisphere between Chimborazo and Loja, whose peak does not exceed that of Saint-Gothard; there are parts of it in the boreal hemisphere, in the Isthmus of Panama, near Cupica, that do not rise above 200 meters (100 toises). But four times the Cordillera attains a colossal mass and elevation. At the 17th degree of austral latitude in Peru, second, at the equator itself in the kingdom of Quito, third, in Mexico at the 19th degree of boreal latitude, and fourth, facing Asia, at the 60th degree of latitude, the height of the peaks exceeds that of the Mont-Blanc, rising to an altitude of 5,000 to 6,000 meters (2,500 to 3,000 toises) of elevation. In general, the Andes chain, even in the high Quito and Mexican plateaus, can astonish our imagination even more by its mass than by its height. On the Antisana volcano at 4,505 meters (2,105 toises) of altitude, I found a plain with a circumference of 12 leagues. The average height of the high Andes near the equator, not taking into account the peaks rising above the ridge, is 3,900 to 4,500 meters (2,000 to 2,300 toises); and the average height of the top ridge of the Alps and the Pyrenees is from 2,500 to 2,700 meters (1,300 to 1,400 toises). The average width of these latter mountain ranges is only 10 to 12 nautical leagues, while that of the Andes is 20 in Quito, and 40 to 60 leagues in Mexico and some parts of Peru. These considerations give a more exact idea of the great differences between the mass of the Andes, the Alps, and the Pyrenees than does the comparison of their highest peaks, which are respectively 6,372 meters (3,270 toises), 4,754 meters (2,440 toises), and 3,434 meters (1,764 toises).

The highest part of the Andes is found between the equator and 1°45' of austral latitude. Only in this small section of the earth does one find mountains exceeding the height of 5,847 meters (3,000 toises). There are only three such peaks: Chimborazo, which would be higher than Etna placed on top of the Canigou peak, or the Saint-Gothard placed on top of the Tenerife peak; Cayambe, and Antisana. The Lican Indians' traditions tell us with some certainty that the Altar Mountain (Altar de los Collanes), which they call Capa-urcu, used to be taller than Chimborazo, but that under the reign of Ouainia-Abomatha, after a continuous eruption lasting for eight years, this volcano collapsed. Thus its summit shows only traces of its destruction on its inclined peaks.

Chimborazo, like the Mont-Blanc, is at one end of a colossal group. From Chimborazo on, up to 120 leagues south of it, no peak is above the limit of perpetual snow. The ridge of the Andes has only 3,100 to 3,500 meters (1,600 to 1,800 toises) of elevation. From the 8th degree of austral latitude, or from the province of Guamachuco on, peaks with snow are more frequent, especially near Cuzco and La Paz, where rise the soaring peaks of Ilimani and Cururana. In Chile, no mountain has been measured, as far as I know; and further south, the Cordillera comes so close to the ocean that the craggy islands of the Huaytecas Archipelago can be regarded as a fragment detached from the Andes chain. The snowy cone of Cuptano, the Teide summit of this region, still rises to 2,900 meters (1,500 toises). But further south, near Cape Pilar, the granitic mountains go as low as 400 meters (200 toises), and sometimes even less. The elevation of the Andes northward from Chimborazo is no less unequal. From the 1°45' of austral latitude to 2° of boreal latitude, the Cordillera maintains an altitude of 5,000 to 5,500 meters (2,600 to 2,800 toises). The Pasto province is one of the highest plateaus of the world: it is the American Tibet. Further north, the Cordillera is divided into three smaller chains. On the easternmost one there are no peaks with snow between the 4th and 10th degrees of latitude: but at its northern extremity, where it bends eastward to form the Caracas mountain range, there is the colossal group of Santa Marta and Mérida; this group is 4,700 to 5,100 meters high (2,400 to 2,600 toises). But the westernmost section of the Andes Cordillera, the one that contains platinum, goes down into the Isthmus of Cupica and Panama, having from 100 to 300 meters (50 to 150 toises) of elevation. When it passes into the Guatemala and Mexico kingdom, from the 11th to the 17th degrees of latitude, its average height is again from 2,700 to 3,500 meters (1,400 to 1,800 toises). But at the 19th degree, in the proximity of the city of Mexico, it forms a group of mountains a few peaks of which, like Popocatépetl and Orizaba, exceed 5,300 meters (2,700 toises) of altitude. In the northern Anahuac and in Nouvelle Biscaye, the Cordillera is barely as high as the Pyrenees. At the 55th degree of boreal latitude, some English travelers even found that it did not exceed 800 meters (400 toises) of elevation. One might be tempted to believe that it disappears entirely toward the north pole, if we did not know that a fourth group exists that is almost as colossal as the others; the Saint Elias peak has an altitude of 5,512 meters (2,829 toises), and Mount Fairweather an altitude of 4,547 meters (2,334 toises). In these areas and in Analasca, the Andes seem to communicate under water with the Kamtchatka volcanoes. The mountains in eastern Asia are only a continuation of the American chain; and as it is probable that the majority of the New Continent's inhabitants belong to the Mongolian race, and as the cradle of the arts, of religious fables, and perhaps of human civilization itself, is to be found in northern Hindustan in the high Tibetan plateau, then it is no less interesting to consider that this same plateau is the common center linking the Cordilleras of the two continents.

I have sketched out here broadly the profile of the high mountain range of the Andes. As far as its structure and the nature of its rocks are concerned, I must limit myself to the following.

The equatorial regions contain all the rocks that have been discovered elsewhere on the earth. The only rock formations that I did not observe there are the steatite rocks, called Topaz rock by Mr. Werner, the mixture of small-grained limestone and serpentine found in Asia Minor, oolite, called *Rogenstein* by the Germans, greywacke, and chalk. But on the surface of the earth, not only are the rocks identical; there is also a harmonious correspondence between the arrangement and the superimposition of these rock masses, which proves that nature acts everywhere according to laws that are as simple as they are universal. In South America, granite constitutes the basis on which the more recent formations lie. It is exposed at the foot of the Andes, on the South Sea coasts, and on the Atlantic Ocean's coasts, between the mouths of the Orinoco and the Amazon Rivers. It supports the elevated structure of the Andes as well as the secondary formations in the plains. The granite, filled with quartz, containing not much mica but large feldspar crystals, seems older in the Andes than the small-grained granite that is filled with little hexagonal tablets of mica. Peruvian granite resembles granite in the high Alps and Madagascar: both are in large masses, sometimes in regularly inclined and parallel layers, with embedded round, very micaceous masses, and both owe their origin to particular attractions between their components. Red titanium oxide is more abundant there than tourmaline. Steatite (*Speckstein*), lepidolite, and barium sulfate have not yet been discovered to be an essential part of granite. On this rock, the most ancient on the planet, and sometimes alternating with it, one finds gneiss, or laminated granite. It yields to micaceous schist, and this in turn yields to primitive schist. In these regions garnet is more commonly found in laminated granite than in micaceous schist. It is even found in the beautiful porphyry that, on top of the schist, crowns the peak of the silver-bearing Potosí mountain. The small-grained limestone rock, chloritic schist, and primitive trap rock (a well-blended mixture of feldspar and amphibole) often occur in subordinate layers in laminated granite and in micaceous schist. The latter is as widespread in the Andes as in the high range of the Alps. It often contains layers of graphite, and is a base for serpentine and jade formations. One sees what has probably not yet been observed in Europe, serpentine alternating with syenite. The high ridge of the Andes is covered everywhere with porphyric formations, with basalts, phonolites, and green rocks. These formations, often divided into separate columns, give to the Cordilleras the aspect of grotesque castle ruins that the spectator first sees when coming upon them from afar. Volcanic fire pierces through these porphyric rocks; and it is a difficult problem for the geologist to solve, whether these porphyries, these basalts, these porous amygdaloids, the obsidians, and the pearly stones were shaped by fire or whether they are preexisting masses on which volcanoes exercise their destructive action.

The similarity of stratification present on the planet's surface is even more striking when one compares the secondary formations of South America with those of the Old Continent. Nature, being characteristically constant, seems to have repeated the same geological phenomena in the plains of the Orinoco, on the coasts of the South Sea, in France, in Poland, and in the deserts of Africa. At the foot of the Andes, one finds two very distinct sandstone formations, one with a silicone base with embedded rocks and sometimes cinnabar and layers of coal; and another formation with a limestone base that glues together secondary rock formations: two gypsum formations, and three secondary limestone formations. Plains measuring over 70,000 square leagues are covered with an old conglomerate containing fossilized wood and brown iron ore. On it rests limestone that can be called the limestone of the high Alps, containing marine petrifications at very high altitudes. Typically it contains many layers of clayey schist and small veins of white limestone spar. It is the base for lamellated gypsum filled with sulfur and often bearing chloride. After this gypsum, one finds another limestone formation, a very homogeneous one, whitish, sometimes full of caves (analogous with the limestone in the Jura, the Monte Baldo, and in Palestine), then one finds a limestone sandstone, then a fibrous gypsum without sodium chloride but mixed with clay, and finally one finds limestone masses containing flintstone and [pierre de corne]. These types of secondary formations are difficult to discern in the immense plains between the Orinoco and the Río Negro, where everything that formerly covered the ancient conglomerate seems to have been carried away by great catastrophes. But it is visible in the New Andalusia province (especially in the Tumiriquiri mountain range) and in Mexico, where Mr. Del Río carried out research that is most valuable for geology. However, despite these similarities in rock formations and stratification in both continents, equatorial regions also have their own particular phenomena. One of the most striking, doubtless, is the immense height reached by rocks that are posterior to granite, and the thickness of these formations. In Europe, high mountains peaks are made of granite. Micaceous schist rarely goes above 2,400 meters (1,200 toises). Granite is found on the Mont-Blanc at 4,754 meters of altitude (2,440 toises). In the Andes Cordillera, this rock is hidden beneath later formations. One could travel for several years in a row in the kingdom of Quito and in parts of Peru without encountering granite. The highest place that I found it in the Andes is in the Andes of Quindiu at 3,500 meters (1,796 toises). The icy summits of Chimborazo, Cayambe, and Antisana, at 6,372 and 5,847 meters of elevation respectively (3,270 and 3,000 toises), are made of porphyry. Secondary limestone goes as high as 3,703 meters (1,900 toises) near Micuipampa in Peru. Huancavelica sandstones are found as high as 4,500 meters (2,310 toises). Micaceous schist in the Tolima Andes is found at 4,482 meters (2,300 toises); basalt in Pichincha, near the city of Quito, is found at 4,735 meters (2,430 toises). The highest point where basalt has been found in Germany is the summit of the Schneekoppe in Silesia at 1,285 meters (660 toises) of altitude. Mineralogists who consider Chimborazo's porphyry, basalts, and green stones not to be altered, but produced by the volcanoes' fire, will also find interesting the research on the limits of these formations' elevations;[21] under consideration here is the state of things as they exist, not their origin nor the primitive state of our planet.

Coal is found in layers near Santa Fé, near the beautiful waterfall of Tequendama at 2,633 meters (1,352 toises). In Peru near Huánuco, it is said that fossil coal has been found in dense limestone rock at 4,482 meters (2,300 toises) of elevation, that is, almost above any current vegetation. The Bogotá plains, at 2,700 meters (1,400 toises) of elevation, are filled with sandstone, gypsum, limestone with shells, and, near Zipaquira, even with rock salt. I do not know if in Europe coal has ever been found above 2,000 meters (1,027 toises) of elevation. What is the cause of this accumulation of the same substances at such different elevations at the equator and in temperate zones?

The highest point where fossilized shells have been found on the Old Continent is the Mont-Perdu, on the highest peak of the Pyrenees at 3,566 meters (1,828 toises) of elevation. In the Andes, debris of organized beings are rather rare in general, because there is very little limestone near the equator. However, near Micuipampa, whose austral latitude I have observed to be 6°45′38″, petrified shells, hearts, Ostrea, and echinites have been found at 3,900 (2,000 toises) of elevation, 200 meters (103 toises) higher than the Tenerife summit. At Huancavelica, some can be found at 4,300 meters (2,207 toises) of elevation.

The bones of fossilized elephants that I brought back from the valley of Mexico, from Soacha near Santa Fé, from Quito, and from Peru, and which Mr. Cuvier determined was a new and very different species from the mammoth, are found in the Andes Cordillera only between 2,300 meters and 2,900 meters (1,181 and 1,489 toises) of elevation. I do not know of any that have

21. *Observations géognostiques faites en Allemagne et en Italie* [Geognostic Observations Made in Germany and Italy, original title: *Geognostische Beobachtungen auf Reisen durch Deutschland und Italien*], 1802, vol. 1, p. 122. This is the work of Mr. de Buch, filled with the most beautiful and most philosophical ideas on the construction of the planet. It would be desirable to have this translated into French.

been found at lower altitudes; the giant bones from the Punta Santa Helena, near Guayaquil, where I did some excavations, were remains of cetaceans.

In Europe, uninterrupted strata with a thickness of over 1,000 meters (514 toises) seem to be very rare. In Mexico and Peru, on the Cordillera's slopes and in some very deep valleys, one easily finds strata of porphyry with a thickness of 3,100 to 3,900 meters (1,600 to 2,000 toises). The porphyry strata of Chimborazo are 3,700 meters thick (1,900 toises). The sandstone near Cuença is 1,560 meters thick (800 toises), and the pure quartz formation west of Cajamarca, which seems specific to the Andes, is 2,900 meters thick (1,500 toises). None of these formations is interrupted by other heterogeneous rocks. Another no less interesting phenomenon characteristic of equatorial regions is that there is a great abundance of porphyry always containing amphibole, but never quartz and rarely mica. The abundant masses of sulfur in the Cordillera are often found far from the volcanoes, not in gypsum nor in limestone mountains but in primary rock strata. I should also mention that the Andes are very rich in all kinds of metals (except lead): I would like to draw the geologists' attention to the *pacos*, that well-blended mixture of clay, iron oxide, silver chloride, and native silver; to the difference in elevation where nature deposited these riches in Peru, at 3,500 meters and 4,100 meters (1,800 and 2,100 toises), and in Mexico, at 1,700 and 2,500 meters (900 and 1,300 toises); finally, to the abundance of mercury, whose veins are innumerable but not mined with great success. But these topics cannot be shown in detail in a general tableau. I will add only one more consideration. Silver is found in such abundance in the Andes Cordillera that Spanish America, which today exports 38 million piasters' worth of silver and gold, will be able to triple its production as its population increases. Today Mexico, where this industry is beginning to exist, exports about 22 to 25 million piasters' worth, compared to 5 to 6 million at the beginning of the eighteenth century. But Europe's wealth did not increase at the same rate; the Mexican currency alone produced, since the conquest, more than 1,900 million piasters, most of which are today in the East Indies and in China.

No region is more wracked by volcanic fire than the Andes Cordillera. From Cape Horn to Mount Saint Elias, there are over 50 volcanoes that still produce flames. The ones furthest from the sea are Popocatépetl in the kingdom of New Spain and Cotopaxi in the Quito province. My observations on longitude indicate that from the Popocatépetl crater to the nearest coast on the Gulf of Mexico (at Tecotutla), the distance is 37 nautical leagues. There are 40 leagues from Cotopaxi to the South Sea. The nature of the volcanoes in the Andes is very different. A few, especially the ones at the lowest altitudes, produce lava: others, for example the ones in Quito, never produce any; but they produce scoria, stones, water, and mostly clay mixed with carbon and sulfur. In a plain in Mexico, during the night of the 14th of September 1759, 29 leagues from the South Sea, the large Jorullo volcano arose from the earth, surrounded by two to three thousand small smoking cones. In a short time, it rose to a height[22] of 486 meters (249 toises) above the former level of the plain. Its elevation above sea level is 1,203 meters (617 toises). It is still active; but we managed, Mr. Bonpland and I, to reach the bottom of its crater to collect air there which contains more than 0.05 carbonic acid.

THE LIMIT OF PERPETUAL SNOW

When considering the decrease in heat in the atmosphere, we saw that, above the Mont-Blanc's height, this decrease seems to follow the same law in temperate zones as in the tropics. One could suppose that in these very high regions the radiating heat given off by the earth's surface becomes almost insignificant and that the temperature of these regions depends almost entirely on the decomposition of the sun's rays in the air that weakens light according to its density. This does not apply in the lower regions of the atmosphere. From sea level to 5,000 meters (2,565 toises) of elevation, the decrease in temperature, taking the average temperature for the whole year, seems to deviate from the law it follows at higher elevations. The atmosphere's strata at which snow does not melt are found at different altitudes according their distance from the pole; but their average temperature must remain the same. Thus, if one knows the decrease in temperature at the equator, from sea level to the limits of perpetual snow, this decrease being 200 meters (103 toises) per centigrade degree, this hypothesis yields the approximate lower limit for snow at other latitudes. One must find the altitude of a stratum of air whose average temperature is +0.4°, the limit above which snow is found at the equator. Let 12.5° be the average temperature of the lower regions at the 45th degree of latitude: then one has $200 (12.5° - 0.4°) = 2,420$ meters; this result corresponds to what is found in nature, give or take

22. The height of this volcano, the most extraordinary and most recent of all, is thus more than three times the height of the large Cheops pyramid in Egypt, which is only 142 meters high (73 toises). It is eight times higher than the Cholula pyramid constructed by the ancient Mexicans out of brick, drawings of which I will publish.

80 to 100 meters (41 to 51 toises). A place in northern Europe having an aver-age temperature at sea level of +4° would have perpetual snow at an altitude of 720 meters (370 toises). In general, this limit, stated in meters, would be found by multiplying by 200 the average temperature of the lower regions. A formula taking latitude into account would be less exact, because physical climate is often very independent from the astronomical position of a place. The same principle that I am presenting here has the advantage of enabling us to determine the average temperature in a country given the altitude of its snow, and of doing this through a multiplier.

But let us set aside hypotheses that are so far based only on a small num-ber of facts, and let us rather see what observation yields. At the equator, the lower limit of snow is one of the most constant phenomena of nature. Bouguer says it is at 4,744 meters (2,434 toises). A large number of mea-surements gave me an average of 4,795 meters (2,460 toises); this difference between Mr. Bouguer's numbers and mine results from the different alti-tudes that Mr. Bouguer and I have given to the Caraburu signal and from the barometer placed at sea level. Scientists have observed quite well that, in a region where the temperature stays the same throughout the year, snow lim-its do not vary by more than 50 to 60 meters (26 to 31 toises) and they form a well-defined horizontal line that does not go into the valleys. The altitude of permanent snow at the 20th degree of boreal latitude had never been deter-mined, and one could suppose that it would be considerably lower away from the equator. Having made geodetic measurements in Mexico, on the Popo-catépetl volcano, on the Itzaccihuatl, on the Orizaba peak, on the Nevado of Toluca, and on the Perote Cofre, I found that perpetual snow begins at 4,600 meters (2,360 toises); the difference with the equator is only 200 meters (103 toises). But some snow falls in Mexico between the 19th and 22nd degrees of latitude, 2,100 meters (1,078 toises) lower than in Quito; this proves that the temporary decreases in the temperature of the atmosphere in these two countries are very different, while their average temperature varies very little. Since the Mexican climate is very similar to that of temperate zones, the perpetual snow limits vary considerably. In July, on the Popocatépetl vol-cano, I found it to be 4,523 meters (2,372 toises); but in February they go as low as 3,824 meters (1,962 toises). The Andes Cordillera does not have any glaciers; this beautiful feature is lacking in this part of the tropics. There is not sufficient snow, which is very rare at the equator, and the temperature is constant: both these factors are probably the reason for the lack of glaciers, whose existence is independent of their altitude. But on Chimborazo, when one digs below the surface, one finds very ancient snows underneath thick beds of sand. We do not know the limits of permanent snow at the 25th and 30th degrees of latitude. In Europe, between the 42nd and 46th degrees, it is at 2,534 meters (1,300 toises) of elevation above sea level. I examined this law of the drop in the limit of snow in a paper read in the first class at the National Institute in the month of Nivôse Year XIII [December 22, 1804 to January 20, 1805].

THE DISTANCE AT WHICH ONE CAN
SEE MOUNTAINS FROM THE SEA

The distance at which one can begin to see a mountain from the sea depends on its height, the curvature of the earth, and terrestrial refraction. The latter being a very variable element, the scale was calculated without taking it into account. But however extraordinary these refraction phenomena might be, one must not forget that, at sea, the uncertainty regarding the place or the position of the ship sometimes led people to believe that they saw objects at much greater distances than they were in reality. The same applies to the effect of currents, the strength of which is often overestimated by the navigator, because he finds himself in a place very far from the one where he thought he was, either through an error or through a lack of astronomical observation.

In the tropics, where I found terrestrial refractions to be extremely constant, the angles of elevation can be of great assistance for the navigator. The Tenerife summit, the one in the Azores, the Orizaba volcano on the Mexican coast, the Silla of Caracas, and the snowy mountains of Santa Marta east of Cartagena in the Indies are signals that nature erected, so to speak, in order to guide pilots. If one knows the height and astronomical position of these summits, one can determine the position of the vessel by means of some very simple observations. Mr. de Churruca calculated tables for the distances at which the Tenerife summit can be seen from various angles of elevation.

The scale I give here also suggests to the imagination the vastness of the terrain that the eye can discover from the high peaks of the Cordilleras. At the point I reached while climbing toward the Chimborazo summit, the breadth seemed to me to be 87 nautical leagues; according to Mr. Gay-Lussac, it was 106 leagues: but clouds and mists hid the lower regions from view for both of us.

THE DIVERSITY OF ANIMALS DEPENDING
ON THE ALTITUDE OF THEIR HABITATS

In order to complete the physical tableau of the equatorial regions, I have
shown in the fourteenth scale the diversity of the animals living at different altitudes in the Andes Cordillera. One finds there, in the interior of the
planet, dermestids that gnaw at underground fungus. The ocean sustains
[bandouillères], coryphenes, and other fish that suck on the gelatinous parts
of fucus. From sea level to 1,000 meters (513 toises) in the regions of palm
trees and scitaminales, one finds sloths living on the *Cecropia peltata*; boas
and crocodiles sleeping at the foot of *Conocarpus* and *Anacardium caracoli*. That
is where *Cavia capybara* hides in marshes covered with *Heliconia* and *Bambusa*,
in order to escape the jaguar's pursuit; *Crax, tanayra*, and parrots perch on
Caryocar and *Lecythis*. That is where one sees *Elater noctilucus* feeding on a sugar
cane, and *Curculio palmarum* living in the pith of coconut trees. The forests in
these torrid regions are alive with the cries of howler monkeys and other
sapajou monkeys. The jaguar, the *Felis concolor*, and the black tiger of the Orinoco, even more bloodthirsty than the jaguar, hunt for little deer (*C. mexicanus*), *Cavia*, and anteaters whose tongue is fixed at the end of the sternum.
The air of these lower-lying regions, especially on the edge of woods and on
river banks, is filled with mosquitoes in such innumerable quantities that
this large and beautiful part of the earth is almost uninhabitable. In addition
to the mosquitoes, there are *Oestrus humanus* that lay their eggs in men's skin
and cause painful swellings; there are acari that attack the skin, venomous
spiders, ants, and termites whose fearful industry destroys structures built by
the inhabitants. Higher up, from 1,000 to 2,000 meters (513 to 1,026 toises),
in the regions of arborescent ferns, there are almost no jaguars, no boas, no
crocodiles, and no manatees, and few monkeys; but there is an abundance of
tapirs, *Sus tajassu*, and *Felis pardalis*. Men, monkeys, and dogs are bothered by
an infinite number of chigoe fleas (*Pulex penetrans*) that are less abundant in
the plains. From 2,000 to 3,000 meters (1,206 to 1,539 toises), in the higher
regions of quinquinas, there are no monkeys, no *Cervus mexicanus*; but there
are *Felis tigrina*, bears, and the great deer of the Andes. At this altitude, the
same as the Canigou summit, lice are unfortunately abundant. From 3,000 to
4,000 meters (1,539 to 2,052 toises), one finds a species of small lion called
puma in the Quichoan language, small bears with white foreheads, and a few
viverrines. I have been surprised to see hummingbirds at elevations similar
to the Tenerife summit. The region of *Espeletia frailexon* and of grasses, from

4,000 to 5,000 meters (2,052 to 2,565 toises) of elevation, is inhabited by groups of vicuñas, *guanaco*, and *alpaca*. Llamas are found only in a domesticated state; those who live on the western slope of Chimborazo became wild when Lican was destroyed by the Inca Tupayupangi. Vicuñas prefer places where snow falls from time to time. Despite persecution, vicuña herds of 300 to 400 animals are still found, especially in the provinces of Pasco, at the sources of the Amazon River, and in the Guailas and Caxatambo provinces, near Gorgor. This animal is also abundant near Huancavelica, near Cuzco, and in the Cochabamba province, near the Río Cocatages valley. It is found wherever the summits of the Andes rise above the height of the Mont-Blanc. It is a very striking aspect of the geography of animals that alpacas, vicuñas, and "guanaco" live all along the ridge of the Andes from Chile up to the 9th degree of austral latitude, but are not seen from that point north, nor in the Quito kingdom, nor in the Andes of New Granada. The ostrich of Buenos Aires presents a similar phenomenon. It is difficult to understand why this bird is not found in the vast plains north of the Chiquitos Cordillera, where thick woods are interspersed with a few savannas. The lower limit of perpetual snow is, so to speak, the upper limit of organized beings. A few lichenous plants vegetate underneath the snow; but the condor (*Vultur gryphus*) is the only animal living in these vast solitudes. We saw it soar at more than 6,500 meters (3,335 toises) of altitude. A few sphinx moths and some flies that we saw at 5,900 meters (3,027 toises) of altitude seemed to me to be carried there involuntarily by ascending wind currents. Mr. Ramond found some around the Mont-Perdu lake. Saussure also saw some on the summit of the Mont-Blanc. I flatter myself that my zoological scale contains the first materials needed for a tableau of a geography of animals similar to the one I executed for plants. Mr. Zimmermann's classic work gives the native habitat of animals according to the different latitudes they inhabit. It would be interesting to establish a profile of the various altitudes where they live at the same latitude.

SOIL CULTIVATION

So far we have analyzed the various physical phenomena of equatorial regions; we examined the changes in the atmosphere, the plants produced by the soil, the animals living at different altitudes, and the nature of the rocks making up the Cordillera. Let us now consider man and the effects of his labor. From sea level to very near the perpetual snow region, our species lives on mountain slopes. The part of Peru called by the Incas Antisuyu, a political division of their empire, is even more populated than Cuntisuyu or the plain. The civilization of peoples is almost always in inverse relation to the fertility of the soil they occupy. The more difficulties nature presents, the more quickly mental faculties are developed. The inhabitants of Anahuac (or of Mexico), those of Cundinamarca (or in the kingdom of Santa Fé), and those in Peru already formed great political associations and enjoyed a civilization similar to that of China and Japan, while men were still wandering, naked and scattered, in the woods covering the plains east of the Andes. But if the civilization of our species makes more rapid progress in the northern regions than amidst the fertility of the tropics, if this civilization began earlier on the high summit of the Cordilleras than on the banks of large rivers, why do already agricultural and civilized peoples not move to climates where nature produces spontaneously everything that must be produced by the hardest labor under a less favorable climate? What makes men till a stony and sterile soil at 3,500 meters (1,796 toises) of elevation, while vast plains below are empty? What makes them live in plateaus where snow falls in every season, and where, under a cold and foggy sky, the soil is devoid of plants? Habit and love of native soil: these are the only reasons that can be given.

In Europe, the highest villages are at 1,600 or 1,900 meters (800–1,000 toises) of elevation above sea level; in the Alps in Switzerland and Savoy, one finds:

	meters	toises
The village of Breuil in the valley of Mount Cervin, at	2007	1030
The village of Saint-Jacques in the Val d'Ayas, at	1631	837
The village of Saint-Remi, at	1604	823
The village of Eleva, on the slope of Cramont, at	1308	672
The village of of Lans-le-Bourg, at	1388	712
The village of Formazza, at	1263	648
In the Pyrenees, one finds, according to Mr. Ramond:		
The village of Heas, at	1465	752
The village of Gavarnie, at	1444	741
The village of Barège, at	1290	662

Higher, in the mountains of Europe, there are only chalets occupied by shepherds during the summer. In the Andes Cordillera, on the contrary, the cities of Pasco, Huancavelica, and Micuipampa are built at almost the same elevation as the Tenerife summit. The Antisana farm in the kingdom of Quito is located at 4,107 meters (2,107 toises) of altitude, and so is probably one of the highest inhabited places on earth.

Soil cultivation depends on the variety of climates, which is the result of its altitude. From sea level to 1,000 meters (513 toises), natives cultivate banana trees, maize, jatropha, and cocoa. This is the region of pineapples, oranges, Mammea, and some very delicious fruit. Europeans introduced sugar, cotton, indigo, and coffee; but these new crops, far from being beneficial, increased the immorality and the misfortune of the human species. The introduction of African slaves, ravaging a part of the Old Continent, brought discord and vengeance to the New Continent.

From 1,000 to 2,000 meters (513 to 1,026 toises), sugar, indigo, banana trees, and Jatropha manihot become scarcer. Coffee prefers a less hot climate, and likes elevated and stony locations. Cotton is still abundant, but not cocoa and indigo which like great heat. In the kingdom of Quito, sugar is cultivated, even quite well, up to 2,533 meters (1,300 toises) of elevation; but then it needs sites where the sun is reflected by vast plains. This temperate region is the most pleasant for European colonizers. They enjoy the temperature of a perpetual spring, and all the fruit, especially those of Annona chylimoya, are most delicious. Above 1,000 meters (513 toises) of altitude European wheat can be cultivated. These nourishing cereals that have accompanied Caucasian peoples for thousands of years can withstand, like man, both the high heat of the tropics and the cold of the peaks near perpetual snow. On the island of Cuba at 23 degrees of latitude, wheat can be cultivated, even abundantly, at 150 meters (77 toises) of elevation above sea level. In the province of Caracas at 10 degrees of latitude, between Turmero and La Victoria, at 500 meters (256 toises), one finds beautiful wheat fields. In the valleys of Aragua, one can be surprised to see sugar, indigo, cocoa, and European wheat cultivated in the same plain. But in order for wheat to grow abundantly in the tropics, in regions so near sea level, there must be an exceptional exposition and a concurrence of circumstances. The real altitude where wheat is produced everywhere is above 1,364 meters (700 toises). In Mexico, for example, in Jalapa, whose latitude I observed to be 19°30′46″, Triticum grows at 1,314 meters (674 toises). It is used to feed livestock, but its ear is almost without seeds. On the eastern slope of the Anáhuac mountains, wheat is grown only from Perote on, above 2,333 meters (1,197 toises). On the western slope, on the contrary, toward the South Sea, I saw wheat growing down into the beautiful Chilpancingo valley at 1,292 meters (663 toises). In other parts of Mexico, as in Peru, in Quito, and in the kingdom of Santa Fé, European wheat grows most abundantly from 1,600 to 1,900 meters (821 to 975 toises) of elevation. There it produces in an average year more than 25 to 30 seeds per stalk. Above 1,750 meters (900 toises), the banana tree does not produce ripe fruit easily; but the plant can grow at 2,500 meters (1,300 toises), though not vigorously. The region between 1,600 and 1,900 meters (821 and 975 toises) is also the one where *Erythroxylum peruvianum* is most abundantly cultivated: this plant is *Coca*, a few leaves of which mixed with caustic lime can sustain the Peruvian Indian during his longest hikes through the Cordillera. From 2,000 to 3,000 meters (1,026 to 1,539 toises) is the elevation where European wheat and *Chenopodium quinoa* are mainly cultivated. Their cultivation is favored by the great plateaus existing in the Andes Cordillera at this elevation, some having from 80 to 100 square leagues. Their soil is even and easy to plow, thus indicating that they are former lake beds. At elevations of 3,100 and 3,300 meters (1,600–1,700 toises), frosts and hail often destroy wheat crops. Maize is scarcely cultivated above 2,339 meters (1,200 toises). From 3,000 to 4,000 meters (1,539 to 2,052 toises), the principal crop is the potato (*Solanum tuberosum*). At around 3,300 meters (1,693 toises), wheat can no longer grow; only barley is sown, and even it suffers from the lack of heat. Above 3,600 meters (1,847 toises), there is no cultivation and no gardening. Men live there among numerous herds of llamas, ewes, and cattle that wander and sometimes get lost in these regions of perpetual snow. The scale of the cultivation of the soil, only sketched out here, gives a picture of the work of man, from the mines up to the highest summits of the Cordilleras.

ELEVATIONS MEASURED IN VARIOUS
PARTS OF THE EARTH

Since all the physical results given in this work were linked to the concept of elevation, it seemed natural to add a certain number of measurements executed in different parts of the planet to serve as a comparison for those carried out in the Andes Cordillera. I have joined them together in the tableau that links the Old and the New Continents, and I have no doubt that these comparisons will suggest some very interesting parallels to those concerned with the vast phenomena of nature.

The picture shows the highest elevations where men have lived from sea level upward. One will see Saussure's voyage on the Mont-Blanc at 4,756 meters (2,440 toises); Bouguer's and La Condamine's on Corazón at 4,814 meters (2,470 toises), and the Chimborazo peak, where we arrived on the 23rd of June 1802 at 5,909 meters (3,032 toises). But all these altitudes seem very small compared to that achieved by Mr. Gay-Lussac, alone in a balloon above

Paris, on the 16th of September 1804. He rose to 7,016 meters (3,600 toises) of elevation, hence almost 600 meters (308 toises) higher than the summit of the tallest mountain on earth. This voyage, which offers a beautiful example of courage and devotion to science, furnished important facts for the theory of magnetism and for our knowledge of the atmosphere's chemical composition.

TABLE OF ALTITUDES

Numbers in parentheses indicate that the measurements are uncertain. The letter H. indicates my own observations, either barometric or geodetic. Some of these will no doubt be changed later in the publication of these measurements and of my astronomical observations; other concerns have prevented me from verifying all of the calculations according to Mr. Laplace's formula, and from giving them the degree of accuracy that will be possible later.

| Places Measured | Height Above Sea Level | | Observers' Names |
	meters	toises	
A. IN AMERICA			
Chimborazo	6544	3358	Humboldt, calculating a part of the height according to Mr. Laplace's barometric formula
	6275	3220	Bouguer, La Condamine
	6587	3380	Don Jorge Juan
Cayambé	5905	3030	Bouguer, La Condamine
	5954	3055	H.
Antisana	5833	2993	H.
	5878	3016	Bouguer
Cotopaxi	5753	2952	Bouguer
Rucu Pichincha	4868	2498	H. (Mr. Laplace's formula)
	4816	2471	Don Jorge Juan
Guagua Pichincha	4740	2432	La Condamine[1]
Tungurahua, after the eruptions of 1772 and the earthquake of 1797	4958	2544	H.
Before these catastrophes	5106	2620	La Condamine

1. The methods used in the barometric calculations also influence this difference that must not be attributed solely to collapses.

TABLE OF ALTITUDES (continued)

| Places Measured | Height Above Sea Level | | Observers' Names |
	meters	toises	
City of Quito[2]	2935	1506	H. (Laplace's form.)
City of Sante Fé of Bogota	2625	1347	H.
Mexico City	2294	1177	H. (Laplace's form.)
City of Popayán	1756	901	H.
City of Cuenca	2514	1290	H.
City of Loja	1960	1006	H.
City of Caxamarca (Peru)	2748	1410	H.
City of Micuipampa (Peru)	3557	1825	H.
City of Caracas	810	416	H.
Antisana farm	4095	2101	H. (Laplace's form.)
Popocatépetl (Mexican volcano)	5387	2764	H.
Itzaccihuatl (Mexican Sierra Nevada)	4796	2461	H.
Sitlaltepetel (Orizaba summit)	5305	2722	H.
Nauvpantepetel (Perote Cofre)	4026	2066	H.
Nevado de Toluca (in Mexico)	4607	2364	H.
Jorullo volcano (emerged from earth in 1759)	1204	618	H.
Mount Saint Elias	5513	2829	Messrs *Quadra* and *Galeano's* expedition
Mount Fairweather (on the northwest coast of America at 60°21′ of boreal latitude)	4549	2334	

2. During the measurement executed in Niguas, Mr. La Condamine found Quito to be 89 meters (51 toises) lower, and this lesser height, like Bouguer's barometric formula and the overestimation of the refraction, modify the altitude assigned to Chimborazo by scientists.

| Places Measured | Height Above Sea Level | | Observers' Names |
	meters	toises	
Arequipa volcano (Peru)	2693	1382	Espinosa
Duida summit (near the sources of the Orinoco)	2551	1309	H.
Silla of Caracas	2564	1316	H.
Tumiriquiri (Mountain of sandstone of the province of New Andalusia)	1902	976	H.
Summits of the Blue Mountains in Jamaica	2218	1138	Edward
B. IN THE SOUTH SEA			
Mowna Roa (Sandwich Islands)	5024	2578	Marchand
C. IN ASIA			
Mount Liban (Tummel-Mezereb summit)	2906	1491	La Billiardière; Icones plant. Syriae, dec. I, p. 5
Ophyr (Island of Sumatra)	3950	2027	Marsden
D. IN AFRICA[3]			
Teide summit	3705	1901	Cordier
	3701	1899	Johnstone
	3689	1893	Borda (Shukburg form.)
	(4313)	(2213)	Feuillé (geodesically)
	(4687)	(2405)	Herberden (geodes.)
	(5180)	(2658)	Man. Hernandez (geod.)
E. IN EUROPE, IN THE ALPS			
Mont-Blanc	4775	2450	Saussure (Shukburg formula)
	4728	2426	Pictet (geodes.)
	4660	2391	Deluc (geod. and barom.)
Mont-Rose	4736	2430	Saussure
Ortler, in Tyrol	4699	2411	somewhat uncertain
Finsterahorn	4362	2238	Tralles

3. Salazes Mountain, on the Island of Réunion, was determined by Lacaille to be 3,300 meters high (1,693 toises); but this measurement is uncertain; the Tafelberg, 1,054 meters high (542 toises).

Places Measured	Height Above Sea Level		Observers' Names
	meters	toises	
Jungfrau	4180	2145	Tralles
Mönch	4114	2111	Tralles
Argentière Peak	4081	2094	Saussure
Schreckhorn	4079	2093	Tralles
Eiger	3983	2044	Tralles
Breithorn	3902	2002	Tralles
Großglockner, in Tyrol	3898	2000	somewhat uncertain
Alt-Els	3713	1905	Tralles
Frau	3699	1898	Tralles
Dru Peak	3794	1947	Saussure
Wetterhorn	3720	1909	Tralles
Doldenhorn	3666	1881	Tralles
Rothorn	2935	1506	Saussure
Le Cramont	2732	1402	Saussure
Wasserberg Rock-salt, in Tyrol	1652	848	Buch
Saint-Maurice Rock-salt, in Savoy	2188	1123	Saussure
Passes that lead from Germany, Switzerland, and France, into Italy:			
Mont-Cervin	3410	1750	Saussure
Seigne Pass	2461	1263	Saussure
Terret Pass	2321	1191	Saussure
Mont-Cenis	2066	1060	Saussure
Petit-Saint-Bernard	2192	1125	Saussure
Grand-Saint-Bernard	2428	1246	Saussure
Simplon	2005	1029	Saussure
Saint-Gothard	2075	1065	Saussure
Splügen	1925	988	Scheuchzer
Les Taures, in the Salzburg area	1559	800	Moll
Brenner, in Tyrol	1420	729	Buch
Col-de-Géant	3426	1758	Saussure
Grimsel	2134	1095	Tralles
Scheidek	1964	1008	Tralles
Pettine, Saint-Gothard Peak	2722	1397	Saussure

| Places Measured | Height Above Sea Level | | Observers' Names |
	meters	toises	
Buet	3075	1578	Saussure
Dôle (in the Jura)	1648	846	Saussure
Montanvert	1859	954	Saussure
Betta Fork	2633	1351	Saussure
Watsmann	2941	1509	Beck
Untersberg	1800	924	Schieg
Hohestaufen	1793	920	Schieg
Rocks of the Pass-Lug	2161	1109	Moll
Schneeberg, near			
Sterzing	2522	1294	Buch
Brenner peak, in Tyrol	2066	1060	Buch
F. NORTH OF THE ALPS, IN GERMANY			
Schneekoppe	1608	825	Gersdorf
Grösse Rad	1512	776	Gersdorf
Tafelfichte	1150	590	Gersdorf
Zobtenberg	721	370	Gersdorf
Hohe Eule	1079	554	Gersdorf
Brocken	1062	545	Deluc
G. IN ITALY			
Etna	3338	1713	Saussure (Shukburg's form.)
Erix Mountain, in Sicily	1187	609	
Monte Vellino			
(Appenines)	2393	1228	Shukburg
Legnone	2806	1440	Pini
Vesuvius	1198	615	Shukburg
Monte-Rotonde			
(Corsica)	2672	1371	Perney
Monte-d'Oro (Corsica)	2652	1361	Perney
Monte-Grosso (Corsica)	2237	1148	Perney
Monte-Cervello			
(Corsica)	1826	937	Perney
Venda (highest peak of			
the Euganean			
mountains)	555	285	Count Sternberg
Monte-Baldo			
(Fenestra peak)	2149	1103	Count Sternberg
Monte-Baldo (peak called			
Monte Maggiore)	2227	1143	Count Sternberg

Places Measured	Height Above Sea Level		Observers' Names
	meters	toises	
H. IN THE PYRENEES			
Mont-Perdu (highest peak of			
the Spanish Pyrenees)	3436	1763	Vidal, Réboul, Ramond
	3366	1727	Méchain
Vignemale (highestpeak of			
the French Pyrenees)	3356	1722	Vidal
Le Cylindre	3332	1710	Vidal and Réboul
Maladette	3255	1670	Cordier (somewhat uncertain)
Le Pic long	3251	1668	Ramond
First Tower of the			
Marboré	3188	1636	Vidal and Réboul
Neouvielle	3155	1619	Ramond
Roland's Pass	2943	1510	Ramond
Pic du Midi	2935	1506	Vidal and Réboul(levelling)
	2865	1470	Méchain (geodetic)
Canigou	2808	1441	Cassini
	2781	1427	Méchain
Bergons Peak	2112	1084	Ramond
Montaigu Peak	2376	1219	Ramond
Passes in the Pyrenees that lead			
from France into Spain:			
Pinède Pass	2516	1291	Ramond
Gavarnie Pass	2331	1196	Ramond
Cavarère Pass	2259	1151	Ramond
Tourmalet Pass	2194	1126	Ramond
I. IN FRANCE			
Mont-Dore	1886	968	Delambre[4]
	2042	1048	Cassini
Cantal	1857	953	Delambre
	1935	993	Cassini
Puy-de-Dôme	1477	758	Delambre
	1592	817	Cassini
Puy-Mary	1658	851	Delambre
	1863	956	Cassini

4. M. Delambre found that Cassini did not take into account terrestrial refraction, so that after recalculation of his observations with the necessary corrections, they are not so far from the truth.

TABLE OF ALTITUDES (continued)

| Places Measured | Height Above Sea Level | | Observers' Names |
	meters	toises	
Col-de-Cabre	1689	867	Delambre
Mezin Mountain			
(Cevennes)	2001	1027	
Le Ballon (Vosges)	1403	720	
Beguines Peak	1115	572	Thuilis and Piston
Sainte-Victoire			
Mountain (near Aix,			
Provence)	970	498	Thuilis
J. IN SPAIN			
S. Ildefonse Palace	1155	593	Thalacker
Picacho de la Veleta			
(Sierre Nevada of			
Grenada)	2249	1154	Thalacker
K. IN SWEDEN			
Kinekulle	306	157	Bergmann
L. IN ICELAND			
Snœfials Sokull	1559	800	Povelsen
Hekla	1013	520	Povelsen
M. IN SPITZBERGEN			
Mount Parnassus	1194	613	Lord Mulgrave

ADDITIONS TO THE GEOGRAPHY OF PLANTS

I

In the course of this work, when some measurements made by Spanish geodesists were used, the reduction of the Castillean vara to meters and toises was not rigorous enough. The vara is to the toise :: 0.513074 : 1.196307, and instead of reducing it by 2.3, one must suppose that the toise = 2.3316 varas. Don Jorge Juan only used 2.32. But see the excellent work by Mr. Gabriel Ciscar, *Sobra los nuevos pesos y medidas decimales* [On the New Weights and Decimal Measures], 1800. The 7,496 varas that the beautiful maps of Depósito hydrográfico in Madrid indicate for Chimborazo are consequently only 3,217 toises, the same number published by Bouguer in his *Figure de la Terre* [Description of the Earth]. Mount Saint Elias has 6,507 varas, or 2,792 toises (5,441 meters). Mount Fairweather has 5,368 varas, or 2,304 toises (4,489 meters). See *Viaje al Estrecho de Fuca hecho por las Goletas Sutil y Mexicana* [Voyage to the Straight of Fuca made by the Schooners Sutil and Mexicana] in 1792; p. cxx, cxv.

II

In 1800 at the Philadelphia Society, Mr. Barton read a paper on the *Geography of Plants of the United States* which is not yet published, but which contains some most interesting ideas. There he states that he found *Mitchella repens* to be the most widespread plant in North America. It is found everywhere from 28° to 69° of boreal latitude. Also *Arbutus uva ursi* is found from New Jersey to the 72nd degree of latitude where Mr. Hearne observed it. On the contrary, *Gordonia francklinii* and *Dionaea muscipula* are confined to a small area. Mr. Barton makes the remark that in general the same species of plants spread further north in countries situated west of the Alleghenies than on the eastern coasts, where the climate is colder. Cotton is grown in Tennessee at a latitude where it is not found in North Carolina. The eastern coasts of Hudson Bay are devoid of vegetation, while the western coasts are covered with it. Mr. Barton observes:

	to the east of the Alleghany	to the west of the Alleghany
Aesculus flava is found from	36° of lat. to	42° lat.
Juglans nigra	41	44
Aristolochia sypho	38	41
Nelumbium luteum	40	44
Gleditsia triacanthos	38	41
Gleditsia monosperma	36	39
Glycine frutescens	36	40

Even *Crotalus horridus* (the rattlesnake) is found east of the Allegheny mountains up to 44°, while west of the mountains it goes as far north as 47° of latitude. See also Mr. Volney's excellent work on the soil and the climate of the United States.

Text of Humboldt's *Tableau physique*

Note: Following is an English translation by Sylvie Romanowski of the complete text of Humboldt's *Tableau physique*. Italicized text indicates notes inserted by the translator or editor. All other text is direct from Humboldt's text. *See table* denotes columns of numerical scales or data, which are not reproduced here. Plant names on the flanks of Chimborazo are not reproduced here.

❶ CAPTIONS AND TEXT OF LARGE PLATE

GEOGRAPHY OF EQUINOCTIAL PLANTS

Physical Tableau of the Andes and Neighboring Countries according to the Observations and Measurements done on Location from the 10th Degree of Boreal Latitude to the 10th Degree of Austral Latitude in 1799, 1800, 1801, 1802, and 1803

BY ALEXANDER VON HUMBOLDT AND AIMÉ BONPLAND

❷ TEXT NOTATIONS ON THE CENTRAL PLATE

To the left of Chimborazo, top to bottom:
Summit of Chimborazo
Point on Chimborazo where Messrs Bonpland, Montufar, and Humboldt carried instruments on June 23, 1802. Bar. 0.3771°, (13 ^{po} n, ^{li} 2) Thermometer −1.6°
Summit of Popocatépetl
Highest point (on Corazón) reached by Messrs Bouguer and de la Condamine in 1738
Height of the Teide summit

To the right of Chimborazo, top to bottom:
Height of 7,016 meters reached by Mr. Gay-Lussac alone in a balloon (above Paris on September 16, 1804), in order to determine the intensity of the magnetic forces, the quantity of oxygen in the air, and the decrease in heat. Barometer 0.3288 meters (barometer in Paris 0.7652 meters). Thermometer − 9.5° (in Paris 30.7°). The intensity of magnetic forces substantially the same as in Paris. Hygrometer of Saussure 33° (in Paris 60°). Oxygen in the atmosphere in the same proportion as at sea level. No hydrogen detected.
Summit of Cotopaxi
Height of the Orizaba summit or Sidatepetel
Height of the Mont-Blanc reached by M. de Saussure in 1787
Height of the city of Quito
Height of Vesuvius

❸ LEFT-HAND PANEL

1st Column: SCALE IN METERS.

See table

2nd Column: REFRACTION at 50° of elevation expressed in seconds of the division by one-hundredth of 0° temperature.

See table

3rd Column: DISTANCE at which mountains are visible from the sea, not taking refraction into account.

See table

4th Column: ELEVATIONS MEASURED in various parts OF THE EARTH.

Elevation of small fleecy clouds
Chimborazo summit, 6,544 meters (3,358 toises), with a reduction of the base at sea level by means of Mr. Laplace's barometric formula.
Cayamba summit   *for numbers, see table*
Antisana summit
Cotopaxi summit
Mount Saint Elias summit
Popocatépetl summit
Orizaba summit
Tungurahua volcano
Rucu-Pichincha summit
Mont-Blanc
Finsterahorn
Fossilized sea shells at Huancavelica
Antisana farm inhabited at
Gross-Glockner (in Tyrol)
City of Micuipampa
Mont-Perdu
Etna
First Tower of Marboré

Watzmann

Canigou

Saint-Gothard

Lower limit [? *unclear*] of snow at 45° of latitude at 2,500 meters of
elevation.

Stratum of rock salt at Saint Maurice in Saneye

Mont-Cenis pass

Mont Dore

City of Popayán

Puy-de-Dôme

Vesuvius 1,198 meters in 1793, but 991 meters in 1805

Brocken in the Harz mountains

Hekla volcano

Kinekulle, one of the high mountains in Sweden

5th *Column:* ELECTRICAL PHENOMENA according to the elevation
of the strata.

Many luminous phenomena.

Few discharges accompanied by thunder. The great dryness of the air
and the proximity of clouds make the presence of electricity very noticeable. Near the mouths of volcanos, it often changes from positive
to negative.

Abundant hail.

Frequent discharges but not regular. Air strata near the ground are
often, and for a long duration, charged negatively. Abundant hail,
even at night.

Above 3,900 meters, hail mixed with snow.

Electrical discharges very frequent and very strong, especially 2 hours
after the sun reaches its highest point. For several hours per day
Volta's electrometer does not indicate 1 millimeter of electricity in
the atmosphere.

6th Column: CULTIVATION OF THE SOIL according to the elevation above sea level.

More cultivation
Llamas, sheep, cows, goats in pastures

Potatoes
Olluco Tropaeolum esculentum
No wheat beyond 3,300 meters
Barley

European wheat. Triticum. Hordeum avena. Chenopodium quinoa. Maize. Potatoes. Cotton. A little sugar. Juglans. Apples (Few African slaves).

Coffee, cotton, sugar cane less abundant. Above 1,750 meters, Musa seldom yields ripe fruit.
Erythroxylum peruvian. Triticum.

Sugar, indigo, cocoa, cotton, corn, Iatropha. Bananas. Vineyards. Achras Mamea, (African slaves introduced by the civilized peoples of Europe).

7th Column: DECREASE in gravity expressed in oscillations of the same pendulum in a vacuum.

See table

8th Column: ASPECT of the azure sky expressed in degrees on the cyanometer.

from 40 to 46°
Average intensity 44°

---

from 32° to 42°
Average intensity 38°

---

from 28° to 37°
Average intensity 32°

---

from 24° to 30°
Average intensity 27°

---

from 17° to 27°
Average intensity 22°

---

from 13 to 23°
Average intensity 18°

9th Column: DECREASE in the humidity of the air expressed in degrees on Saussure's hygrometer.

No observations

The average dryness of the air without clouds is probably above 38°, which reduced at a temperature of 25.3° is probably 26.7.°

From 46° to 100°
Average humidity 54°

From 51° to 100°
Average humidity 65°

From 54° to 100°
Average humidity 74°

From 60° to 100°
Average humidity 80°

From 65° to 100°
Average humidity 86°
Rainfall averages 1.89 meters (70 inches), in Europe 0.67 meters (25 inches).

All these hygrometric measurements are affected by the average temperature indicated on the temperature scale.

10th Column: Atmospheric air PRESSURE expressed in barometric levels.

See table

11th Column: SCALE IN TOISES.

See table

❹ RIGHT-HAND PANEL

1st Column: SCALE IN METERS.

See table

2nd Column: TEMPERATURE of the air at various heights, maximum and minimum on the centigrade thermometer.

Regions too little known to determine the average temperature, which however appears to be below zero. At 5,403 meters, the thermometer sometimes climbs to 17.8°.

---

from −7.5° to 18.7°
Average temperature 3.7° (3° R)
Snow falls till 4,100 meters of elevation.

---

from 0° to 20°
Average temperature 9° (7.2° R).
Abundant hail, sometimes even at night

---

from 1.2° to 23.7°
Average temperature 18.7° (15° R)
Very abundant hail. Frequent mist at not very high elevations.

From 12.5° to 30°
Average temperature 21.2° (17° R).
Hail rather rare. Sky frequently overcast.

From 18.5° to 38.4°

Average temperature 25.3° (20.2° R).

No hail. Sand frequently at 52°.

Water temperature at the surface of the sea near the equator, outside the currents, 28°, but at depths of 400 meters the sea is 7.6°. The temperature of the interior of the earth seems to be 22.3° at the equator.

3rd Column: CHEMICAL COMPOSITION of the atmosphere.

The quantity of oxygen in the atmosphere appears to be the same in the elevated regions and in the plains. But the proximity of volcanos can sometimes modify the composition of the air in the vicinity of the high peaks of the Andes.

The quantity of hydrogen contained in the air is decreased by two-thousandths. More hydrogen is not found above 7,000 meters of elevation than at sea level.

Air contains 0.210 of oxygen, 0.787 nitrogen, and about 0.003 carbonic acid. The greatest variations do not seem to exceed one-thousandth of oxygen.

4th Column: ELEVATION of the lowest limit of perpetual snow at various latitudes.

Air in water melting from snow contains 0.287 oxygen.

Perpetual snow at the equator and from 3° of boreal latitude to 3° of austral latitude, 4,800 meters (2,464 toises).

No variation in 80 meters of perpetual snow at 20° of boreal latitude at 4,600 meters (2,361 toises) but in winter it goes down to 3,800 meters.

Perpetual snow at 35° latitude at 3,500 meters (1,800 toises) of elevation.

Perpetual snow at 40° latitude at 3,100 meters (1,600 toises) of elevation.

Perpetual snow at 45° boreal latitude at 2,500 meters (1,282 toises) of elevation.

In the Pyrenees at 2,440 meters. In Switzerland at 2,700 meters on isolated summits; at 2,530 meters if the mountain summit is above 3,100 meters. Phenomena not constant in variable zones.

At the equator snow falls at 4,100 meters (2,100 toises) of elevation. In Mexico at 19° of latitude it falls at elevations up to 1,800 meters.

Perpetual snow at 75° of boreal latitude.

5th *Column*: SCALE of animals according to the elevation where they live. No organisms living fixed in the soil.

The condor of the Andes, a few flies and sphingids flying about in the air, perhaps brought to these regions by ascending air currents.

Vicuñas of Guanaco, alpacas in numerous groups, a few bears, condors, falcons Caprimulgus. Many fish in the lakes.

Llamas turned wild on the western slope of Chimborazo. The little bear with a white forehead. Large deer. The small lion. A few hummingbirds. More Pulex penetrans.

Viverra mapurito. Felix tigrina. Large deer. Palamedea bispinosa. Abundance of ducks and diving birds. Many lice (Ped. hum.).

Small deer (Cerv. mexic.). Tapir. Sus Tayassa. Felix pardalis. A few howler monkeys. Troupial (Oriolus). Coluber coccin. No boas, no crocodiles, many jiggers (Pul. penetr.).

Many sapajou and howler monkeys. Jaguars (Felix onca). Black tigers. Lions (Felix concolor). Cavia capibara. Sloths. Anteaters. Cerv. mexic. Armadillos. Penguins. Crax Ampelis. Boas. Crocodiles. Manatees. Blater noctil. Mosquitoes (Oestr. human.).

In the interior of the earth, new species of dermestid beetles chew on subterranean plants.

6th Column: TEMPERATURE at which water boils at various altitudes. Centigrade thermometer.

See table

7th Column: GEOLOGICAL ASPECTS.

The nature of rocks appears in general to be independent of differences in latitude and in height. But when one considers only a small part of the earth, one finds that in each region a particular system of forces determined the order of the rocks' stratification, their inclination, and the orientation of their strata. It is recognized that certain local laws determine the elevations of the various formations above sea level.

Equatorial regions contain both the highest peaks and the widest plains on earth, from 0° to 1°45', and nowhere else on earth does the height of mountains exceed 5,850 meters. But towards the poles, however, mountains do not get very much lower, for one finds at 19°, 45° and 60° of boreal latitude some peaks with 4,700 and even 5,500 meters of elevation.

On the contrary, equatorial plains, those between the easterly slope of the Andes and the coasts of Brazil, do not have more than 70 meters to 200 meters of elevation above sea level.

All the structures that have been found on the rest of the earth are found together at the equator. Their relative age, which is visible in the order of their stratification, seems in general to be similar to that in temperate zones. Granites form a base for gneiss, syenite and micaceous schist, and primitive schist; all the secondary formations, two of sandstone, two of gypsum, and three of limestone, are a striking example of the identity of structures in the most distant parts of the earth. The problematic formation of the basalts, of the amygdaloidal basalt, of amphibolites, and of porphyry with obsidian and pearly stone (Perlstein) is as sparse in the high ranges of the Andes as it is in the high mountain ranges of Europe.

Among the features particular to the equatorial regions of the New Continent, one must mention especially the thickness of the strata and the great elevation where one finds formations posterior to granite. In Europe, granite is not covered by other rocks from 3,300 to 4,700 meters. In the Andes, granite is not visible above 3,500 meters. The

highest peaks on earth are composed of porphyry in which am-
phibolite is abundant, containing no quartz, and considered by some
mineralogists to be produced, by others to be altered, by volcanic fire.
Sandstone formations are found in Huancavelica at 4,500 meters.
Coal is found at Huanuco at an elevation of 4,400 meters. The plains
of Bogotá at 2,700 and 2,900 meters are covered with sandstone,
secondary limestone, gypsum, and rock salt. Fossilized sea shells are
found in the Andes above 4,200 meters (in Europe, they have not been
found above 3,500 meters).

The soil in the kingdom of Quito contains, at 2,500 meters of elevation,
huge deposits of elephants' bones, from a species that appears to
have been destroyed.

The sandstones at Cuenca are 1,560 meters thick; a quartz formation
west of Caxamarca is 2,900 meters thick.

The Andes Cordillera has more than 50 active volcanoes, some of which
are 37 to 40 nautical leagues from the sea; the highest ones and the
ones with the most reinforced sides do not emit scoria, and especially that carbonated earth in which is often found a fish (Pimelodus
cyclopum).

8th Column: INTENSITY of light in the air at various elevations, measured
in units of intensity of light in a vacuum.

See table

9th Column: SCALE IN TOISES.

See table

Humboldt's Pictorial Science:
An Analysis of the Tableau physique des Andes et pays voisins

Sylvie Romanowski

> ... I had conceived an idea of my own, namely that the important thing was
> to form some visual images, and the eyes would come later in consequence. So I
> concentrated on making the part of me that was outside (and even the interior part
> of me that conditioned the exterior) give rise to an image, or rather to what would
> later be called a lovely image . . .

ITALO CALVINO, Cosmicomics

INTRODUCTION: AN OVERVIEW OF THE VOLUME

The principal title page hints at the variety and the scope of the texts to come:
the general title, The Voyage of Humboldt and Bonpland, announces the first volume containing the Essay on the Geography of Plants accompanied by a Physical
Tableau, and adds simply, "With one Plate." With this plate, however, Humboldt invents a complex mode of visualizing scientific data, corresponding
to his new science of the geography of plants, itself a part of his idea of a
future general physics.

The plate is folded into a well-organized volume consisting of three texts
of increasing lengths, a "Preface" to the "Essay," the "Essay," and the "Physical Tableau." Preceding the "Preface" is a page in very ornamented script
(the only example of such ornate lettering in the entire volume) dedicating
the work to two eminent personalities, Jussieu and Desfontaines, who are
not mentioned again in the rest of the volume.[1] The "Preface" is a succinct

---

Note: All French and English spelling has been modernized.

1. Dedications to important people, such as royal or aristocratic patrons, were nor-
mally used in most works, both fiction and nonfiction, of the ancien régime, and were
part of the signification that was intended for the work by the author. By the eighteenth
century, as authors became more independent from patronage, dedicatory practices

text in which Humboldt expresses the purpose of the work, his intention to continue this kind of research and writing, and his gratitude toward his contributors and predecessors and his Spanish hosts. The "Essay" (pp. 64–75) states his main argument and his polemical proposal to establish a new science of the geography of plants. He makes several pleas. The first is for eschewing the study of individual plants and deemphasizing classification in favor of the study of plant societies. He then pleads for a study of nature in its totality, stating that nature should be studied not in hothouses but on the terrain [dans la nature même] (p. 74), and that this new approach should inspire artists to feel and paint its beauty. The "Essay" concludes with a reflection on the moral value of studying the laws of nature, a study that gives humankind a strong sense of pleasure and freedom. Let us note that the word "tableau" appears only once in this "Essay," in the esthetic sense of a work of art done by a painter.

The "Physical Tableau of Equatorial Regions," the last and longest section, is organized in five parts. The first three parts are structured by a movement of ascent from sea level to the mountain peaks, as stated in the first sentence. The first section (pp. 76–78) gives a thesis statement in four long paragraphs: plants must be studied according to their situation in altitude and latitude. From p. 78 (starting with "I have attempted to gather in one single tableau") to p. 87, Humboldt mentions the *Tableau physique* itself for the first time in the work. Here, as I will discuss later, the word *tableau* refers in an ambivalent fashion to the text and to the plate. Regarding the plate, he discusses "two conflicting interests, appearance and exactitude" (p. 81). He explains the difficulties encountered in representing the mountains as they are in reality, the impossibility of being true to scale on a limited length of paper, and how he placed plant names according to their altitude. In a third section, from "In order to bring together" (p. 87), he lists in detail the plants he saw and gathered, again organizing his account in an ascent from underground plants to the "last organized beings we found in the ground at these great heights" (p. 94). In a fourth section (pp. 94–99) he broadens his perspective horizontally, advocating that this same kind of tableau be carried out in other parts of the world, for temperate regions in both the northern and southern hemispheres, for Africa and Asia. He concludes with the statement that his tableau of equinoctial regions is particularly condu-

---

gradually waned, but they did not disappear entirely. A vestigial form can be seen here in Humboldt's naming of two eminent scientists.

cive to giving us an overview of all our knowledge about how phenomena vary in relation to altitude. The fifth and longest section, from p. 100 on, is a detailed description of sixteen of the twenty scales on either side of the plate (excluding the four scales showing meters and toises). Lastly, two appendices give, first, tables of numbers of comparative heights of mountains in the world, and second, two notes, one on his use of Spanish data and the other on some observations parallel to his that were done in the United States (the latter being the very last words of the volume).

The last section on the scales (pp. 100–136) merits a closer look, as it embodies Humboldt's science in its two main aspects. Measurements are at the heart of his science, and, equally strongly, high altitudes, which clearly fascinate him. In this section, we find a strong sense of upward movement. The discussions of most of the scales start with words like "when one rises above sea level," and all mention the differences between phenomena at sea level (and a few times, even further below in the interior of the earth), and the highest mountain peaks.[2] And it is in this part alone that mention is made, no less than sixteen times, of Gay-Lussac (1778–1850), who, in his balloon ascents, reached the highest altitudes a human being could at that time. The tone of admiration is palpable in the last sentences: "But all these altitudes seem very small compared to that achieved by Mr. Gay-Lussac, alone in a balloon above Paris, on the 16th of September 1804. He rose to 7,016 meters (3,600 toises) of elevation, hence almost 600 meters (308 toises) higher than the summit of the tallest mountain on earth" (p. 136). The concluding sentence of the essay brings together science, moral strength, and the fascination with the altitude achieved by Gay-Lussac: "This voyage, which offers a beautiful example of courage and devotion to science, furnished important facts for the theory of magnetism and for our knowledge of the atmosphere's chemical composition" (p. 136). This is the true and fitting conclusion of the overall work, extolling both personal courage and the devotion to science that are inseparable in both Humboldt's life and in his contemporary's exploration of the atmosphere.

As this brief overview has shown, the work is composed of several distinct parts ranging from the polemical to the purely scientific. The complexity of

2. It is interesting to note that in writing *Cosmos*, Humboldt also organizes his text with respect to altitude, but in the opposite direction, in a descent: "Beginning with the depths of space and the regions of the remotest nebulae, we will gradually descend through the starry zone . . . to our own terrestrial spheroid" (79).

this book exemplifies Humboldt's approach to scientific knowledge and to the communication of science at this time (later he will instead write long, encyclopedic works, but with the same goal): the simultaneous layering of many aspects of the vast subject matter into one work, here a short one, later, much longer works, an approach that is equally present in the plate.

In this essay, after examining the relation between the prose texts and the illustration, I will analyze the plate itself as an esthetic and visual artifact, drawing on present-day scholarship on graphics. The production of such a complex plate is embedded in its historical and philosophical contexts of the period. In this period, many significant scientific illustrations were devised and refined, some of which have a bearing on Humboldt's plate. Humboldt must also be situated in the context of the latter part of the Enlightenment, especially the French Enlightenment, when there were important tensions in scientific and philosophical thought regarding scientific systems, particularly at the time that the Enlightenment was yielding to Romanticism. I will conclude with some remarks on the significance of Humboldt's use of a plate to convey his new vision of and for science.

TWO TABLEAUX, ONE NATURE: TEXTS AND PICTURE

The word *tableau*, which I have kept in the translation, is in the title of the plate, and it appears no less than thirty-five times in the text, "The Physical Tableau of Equatorial Regions." But what does *tableau* refer to exactly?

The word *tableau* applies clearly both to the text and to the plate. When Humboldt says in the "Preface" that if he could have worked more on this book [cet ouvrage], it would have been shorter because "a tableau must contain only the general physical qualities" (p. 61), the mention of the length of a work refers to a text, not to a picture. When he enumerates the many details contained in the "Physical Tableau" (p. 76 on), he is referring to his picture as well as to his text, which obviously can contain many more words than he can put in a picture. He regrets the constraints of having to put all the information he gathered in the Andes in a single work, which could refer to the text or the picture: "I have attempted to gather in one tableau the sum of the physical phenomena present in equinoctial regions" (p. 78). He is referring clearly to the picture two pages later when he discusses the differences between it and "botanical maps" (p. 80). The reference of the word *tableau* is ambivalent, as it refers most of the times to the picture, but sometimes to the text. The ambiguity is even greater in French, where *tableau* can mean what is covered in English by the words *board, picture, table,* and *list.*

This suggests that the words and the picture are inseparable—the two together form one work, and only combined into one work can they convey Humboldt's knowledge. The illustration is an adjunct to the text, but the text is one long caption to the illustration. What we have here is a plate accompanying a text, but also a text accompanying a picture.

However, we have not one but two texts, the "Essay" and the "Physical Tableau." The short polemical "Essay" affirms Humboldt's rejection of the narrow-minded botanists concerned only with discovering new species of plants and studying their form and their classification independently of their environment. What Humboldt is aiming at is a new holistic science that he calls "general physics" [la physique générale] (p. 64). His approach to the study of nature is conveyed in the "Essay," where the word *tableau* is used only once; there it refers not to the plate but to an imagined painting (p. 74), in a discussion of the esthetic value of the spectacle of nature as a whole. The "Essay" is followed by the longer "Physical Tableau," which explains in detail the content of sixteen of the twenty columns framing the profiles of the mountains. In other words, the "Essay" functions as a caption for the "Physical Tableau," and in turn both prose works together form a caption for the pictorial *Tableau physique*, and all three works together produce a spectacle of nature that is philosophical in its intent, scientific in its method, and esthetic in its impact.

The prose texts are obviously essential to Humboldt's purpose, which could not be carried out by a picture alone. Yet he finds the plate equally necessary, hoping that it will enhance the appeal of his new approach to the study of nature: "I thought that if my tableau were capable of suggesting unexpected analogies to those who will study its details, it would also be capable of speaking to the imagination" (p. 79). I will return later to the notion of analogy as central to his method of communication. Humboldt, a stickler for precise measuring, is aware of the tension between "two conflicting interests, appearance [l'effet] and exactitude" (p. 81). Nevertheless, he is willing to sacrifice some scientific exactitude to produce this picture, because it will appeal to the imagination, thus showing his belief that imagination is essential to the intellectual work of science. He finds it constraining to include all the details "within the limits of a single plate" (p. 86). On one sheet of paper, the scale of horizontal distance and the scale of height cannot be the same, and the latter, in keeping with his constant attention to altitude, has precedence: "one would need a sheet of paper 40 times as long as the format of this book" (p. 85). The pitch of the slopes is distorted (p. 85), and some explanations are needed about the relationship of the two mountains depicted: Chimborazo and Cotopaxi are not as near to each other in nature as they appear on paper, but Cotopaxi is much more interesting to show because it is still an active volcano (indicated by the plume of smoke rising above it), and because Humboldt himself heard its rumblings from as far away as Guayaquil, while the mountain next to Chimborazo is actually Cargueirazo, which unfortunately collapsed in 1698. Not only is the picture essential to his purpose despite its shortcomings, but these very problems can themselves be interesting and are to be the topic of a future work (p. 85). The duality of this volume forces the public to both read and look, to be engaged intellectually and imaginatively with ideas and spectacle. Humboldt's epistemology is no less holistic than his science.

THE PICTURE AS ESTHETIC ARTIFACT

What makes this picture so essential to Humboldt's purpose? I would like to look at it in detail, as a freestanding pictorial representation.[3] It is a structure, in the modern, structuralist sense of the word—so that each element is dependent on every other element and is inseparable from the whole.[4]

At first glance, the plate seems relatively straightforward, utilizing a shape well known since medieval times: it is a horizontal, rectangular triptych with a central panel showing a profile of two mountains, Chimborazo with Cotopaxi behind it, on a background of blue sky, flanked by two panels consisting of vertical columns of numbers and words. What makes the picture interesting is the organization and the subtle variations among its elements. In the central panel, the mountains are divided in proportions of about one-third green foliage to two-thirds white covered with words. One part of Chimborazo, with its distinctive cone, is more realistic from base to summit, while the other part is itself divided by a crevasse and almost entirely covered with words, except for the peak behind the text with its plume of smoke and for a profile of grass on the lower right-hand slope. Yet the realism of Chimborazo is muted, as the picture makes no attempt to depict three-dimensional space; the more realistic part of Chimborazo is just as flat as the word-laden

3. The plate is discussed briefly in Nicolson 1990; Dettelbach 1996a, 268–70; Godlewska 1999b, 261–63; Bowen 1981, 222–25; Rupke 2001, 99–101; Tresch (forthcoming). I would like to thank John Tresch for allowing me to read his article in manuscript.

4. "One seeks to explain the particularities of a literary work by the relations existing among its constitutive elements, or the relations that the work has with other works" (freely translated from Todorov 1968, 99–100).

part. This could give the profile a curious impression of floating in space, but it is anchored by a clear, horizontal line separating the green vegetation and the tapestry of words from the blank space with only a few words on it that represents the ground below sea level; on either side of the mountain profiles, the sea level is briefly but clearly indicated with a bit of blue color. The mass of the mountain is balanced by a large expanse of sky, a negative space that is white a third of the way up the profile, shading gradually from white to deeper blue at the top. The representation is flattened, devoid of depth, and lacking any dynamic diagonal thrusts that painters often employ to make their paintings dramatic. There is no focal or vanishing point, and thus the eye can keep moving: sideways from green vegetation to words on white, from mountain to sky, up and down from white to blue, and across the sky along the thin engraved lines in the background indicating cloud layers. The sky space is not covered with words like the mountains, but is not totally blank either, since it contains quite a few pieces of narrative and information: the names of the mountains themselves, the names of other mountains at their respective heights, on the right a paragraph on the altitude reached by Gay-Lussac in a balloon ascent over Paris in 1804, on the left several sentences on Humboldt's and Bonpland's climb of Chimborazo and an ascent by Bouguer and La Condamine of El Corazón in 1738.

Thus the picture is composed of sections balancing each other, words and space, sky and earth; colors of blue, white, and green are balanced by the deepest color of green that is off center in the lower bottom quarter of the picture. In spite of its clearly flattened, two-dimensional aspect, the picture is not static. It makes the eye circulate, as if to say: The spectacle of nature is one that involves movement, elegant variation, and well-balanced proportions. Nature can be understood, represented, and mastered, and it demands constant movement of the eye and intellect. The intellect and the eye are also drawn to the columns framing the central panel. They provide a strong contrast with the profile, which thereby appears much more representational than it would by itself. Together the columns make up about half of the total space of the plate and hence are hardly mere marginal adjuncts.[5] The two sets of columns are about the same width, and each set is framed by the left-most

5. Dettelbach 1996a emphasizes, more than I do, the marginality of the columns and the conflict that Humboldt felt over the inclusion of the margins and their data: "the inscription of unpicturesque Latin names in the right half of the mountain . . . marks a further compromise, likewise regretted" (271). At this point, Humboldt is confident or bold enough to try to represent everything at once. Later, he was to simplify his plates, as can be seen in his next work, the *Atlas* (1812).

and right-most columns consisting only of numbers indicating altitude, meters on the left and toises on the right of each set. But there are variations:
eleven columns on the left, nine columns on the right. The columns on the
left are slightly more numerous but have briefer bits of text, while the ones
on the right are fewer but one of them contains much more text. And there
is a general pattern: in both sets the text in the columns rises progressively
higher from left to right. The eye also moves from columns with less text to
the text-rich column on the right labeled "Geological Considerations," so
that the eye, in the habitual movement of reading from left to right, encounters progressively more text, though not in a steadily mechanical increase.[6]
The linear verticality of the columns filled with words and numbers contrasts
with the natural shapes of the profile, and the movement of the eye going up
and down each column is balanced by indications that the eye is also invited
to move horizontally from column to column, and from columns to picture,
guided by the horizontal background lines, the layers of the clouds, the lines
of prose on the sky, and the strong demarcation at sea level at the foot of the
mountains.[7] This plate is a rich, harmonious whole, with variations among
all its elements, forming a well-defined structure that encompasses pictorial
representation, text, and numbers working together.[8]

The final element of the plate is the lower third of the page, containing
the title and information on its content—and here too there is a balance

6. The columns are analyzed here from an esthetic point of view: there is a pattern
of diagonally rising text. This analysis contrasts with the one done in this volume from
a purely scientific point of view by Stephen Jackson, who says that the columns "do not
seem to be in any particular order" (p. 24). Since presumably there is no scientific principle that is violated by the randomness of the order, Humboldt is free to pay attention
to giving the columns a pleasing organization.

7. For two visual experiments, first, to see the effect of looking at the columns by
themselves, see Dettelbach 1996b, 292–93. Here the columns are devoid of reference,
like sign-posts to nothing. And second, to see the picture without the columns, see
Bowen 1981, 224: here the picture seems emptied of any frame, much less informative
and scientific. Together these illustrations provide another demonstration of the unity
of the plate.

8. Here I disagree with Dettelbach 1996a, 271, who states in his otherwise suggestive
analysis that Humboldt is torn between the demands of precision and picturesqueness.
Humboldt wants to give equal importance to both. His later description of nature's
variations is apt here: he wants to show not "the peaceful charm uniformly spread over
nature" but rather "the peculiar physiognomy and conformation of the land, the features of the landscape, the ever-varying outline of the clouds, and their blending with
the horizon of the sea" (Cosmos 26).

between the elegantly shaded capital letters of the main title, the italic script of the remainder of the title, and the print size that decreases progressively toward the bottom yet remains completely legible. The next to bottom line is again in vertical capitals, a bit bigger than the line immediately below, and contains the names of Humboldt and Bonpland—smaller than the main title but easy enough to read, as if the authors want to name themselves clearly but not attract too much attention. All these elements are in equilibrium and derive significance from their juxtaposition on the page: the mountains' profiles, the columns of numbers and words, the sky, and the bottom section of the page, all convey different information by being next to each other and by inviting the reader/spectator to continually pass from one element to the next.

How can one evaluate in a precise manner the effectiveness of the plate in conveying information? Much critical attention has been paid recently to the techniques of scientific illustration and the use of visualization in science, in particular by Edward R. Tufte and Howard Wainer, whose work can be useful in evaluating Humboldt's plate. Tufte asks a succinct question, which could also sum up Humboldt's problem: "The world is complex, dynamic, multidimensional; the paper is static, flat. How are we to represent the rich visual world of experience and measurement on mere flatland?" (Tufte 1990, 10). In an earlier work, Tufte poses the problem thus: "how to communicate information through the simultaneous presentation of words, numbers, and pictures" (Tufte 1982, 10). In his most recent work, Tufte (2006, 122–36) has offered six "fundamental principles" of analytical designs: comparison, causality, multivariate analysis, integration of evidence, documentation, and content, which "counts most of all" (Tufte 2006, 134).

Comparisons are certainly omnipresent in the picture: Humboldt includes many points of reference, both from the Andes region and from Europe, to stress the immensity of the mountain ranges he encountered in South America. Beside Chimborazo the European Alps seem puny indeed, but the achievements of scientist-explorers like La Condamine, Gay-Lussac, Bouguer, and Humboldt and Bonpland themselves, who are modestly named at the bottom of the plate, appear all the greater. In many other profiles drawn by Humboldt, Chimborazo, on which he climbed without the benefit of modern technology as high as he could till he suffered too much, is shown as a point of reference and testifies to his own exploit—Chimborazo was to remain associated with Humboldt and appears in many of his later portraits. The second criterion listed by Tufte, causality, is one that no doubt Humboldt would have liked to consider, but it is difficult to see how he could have adapted the plate to represent cause and effect.

Multivariate analysis is one of Humboldt's main concerns: the numerous aspects of the nature he explored are obviously enumerated in the columns and inscribed on and around the mountains. But the integration of evidence is more questionable. Certainly the plate shows many variables, but are the words, numbers, images, and diagrams well integrated? Or do they coexist side by side, forcing the readers to integrate all the variables themselves? Perhaps the very plethora of details works against a clear view of what Humboldt wants to convey. One critic, with some justification, has noted the "clutter of textual information" on Humboldt's plate (Rupke 2001, 100)—one part of the plate is indeed cluttered, as the names are not printed on straight horizontal lines but curve in different directions and hence appear a bit jumbled, and the columns can also appear rather full of details. But seen as a whole the plate does not appear unharmonious. In his next major graphic work, the *Atlas géographique et physique du royaume de la Nouvelle-Espagne* (Geographic and Physical Atlas of the Kingdom of New Spain) (1811) accompanying the *Essai politique sur le royaume de la Nouvelle-Espagne* (Political Essay on the Kingdom of New Spain) (1811), Humboldt does not attempt such complexity: none of the plates exhibit the same sheer quantity of information. They contain only one vertical scale of height, few words, one cross-section at a time without the telescoping of mountains, one color, a dark brown with no attempt at realism. In other words, the 1811 profiles are less cluttered, simpler, easier to read, but also less complex and more analytic than synthetic.[9] However, the large quantity of information on the *Tableau physique* never becomes what Tufte calls "chartjunk," useless details that fill up space without conveying meaning and that obfuscate the information. There is no useless decoration on this plate. That Humboldt was eager to avoid any detail that might seem merely decorative is indicated in his scrupulous explanation of why he added a plume of smoke to a volcano in one of his illustrations in the *Atlas de la Nouvelle-Espagne*, an explanation which would apply here too: "In order to distinguish the active from the inactive volcanoes, I allowed myself to add a

9. To see what these profiles look like, see Dettelbach 1996a, 265, fig. 12.3, which reproduces two profiles of the Iberian peninsula, which look very similar to the profiles of the *Atlas of new Spain*; Godlewska 1999a, 262, fig. 8.5 reproduces part of a plate from the *Atlas*. For a similar illustration, see fig. 3.

small column of smoke on the drawings of Orizaba and the great volcano of Puebla, even though I myself never saw this smoke either at Jalappa or even in Mexico" (in his *Analyse raisonnée de l'Atlas de la Nouvelle-Espagne* (Systematic Analysis of the Atlas of New Spain), which precedes the *Essai politique* itself, commentary on Plate XVII). Humboldt's simplification of his subsequent maps indicates perhaps that he was aware that the data must not overshadow the representation, mindful of a principle as stated by Wainer: "A graph that calls attention to itself pictorially is almost surely a failure" (Wainer 1997, 11).

A fifth criterion is what Tufte calls documentation (2006, 132–33), that is, details such as the identification of the authors, printers, and sponsors, the data sources, scales, and assumptions: these details are all meticulously indicated, down to the fine print at the very bottom of the page listing the exact contributions of the principal people concerned: "Sketched and written by Mr. von Humboldt, drawn by Schönberger and Turpin in Paris in 1805, engraved by Bouquet, lettering by Beaublé, printed by Langlois."

This brings us to the discussion of the content: the quality, relevance, and integrity of the information on this plate. If the quantity of information on the plate is large, even a bit crowded both on the profile and in the columns, the quality of its content is the heart of the matter, and the painstaking information Humboldt gives in his text and on the plate leaves no doubt how demanding he was of himself and Bonpland regarding the quality of their observations. The details regarding which instruments and scales were used, the conditions in which measurements were made, all point to their exactitude of information-gathering. As much information is given in the columns as possible, to be completed by the prose "Physical Tableau" which is, as I mentioned, a long legend explaining the information included on the plate in all three panels.

The plate is a remarkable work of both science and art that stands up to a critique informed by modern experts in graphics who come much later and have seen many illustrations both good and bad. Howard Wainer, speaking of graphs in earlier periods, has observed: "The graphical presentation of scientific phenomena served two purposes. Its primary function was standardizing phenomena in visual form, but it also served the cause of publicity for the scientific community" (Wainer 2005, 6). Certainly, Humboldt's plate addresses both these aims, which are clearly stated in the "Essay." The preoccupation with observation, comparison, and standardization, manifested here in the juxtaposition of numerous phenomena found at diverse altitudes, many of which were never before observed or measured firsthand at such heights, is the scientific goal, but the publicity goal is also equally present. In the "Preface," Humboldt states that he would like to repay the public's interest in his expedition with his publication, but first he finds it appropriate to "draw the physicists' attention to the broader phenomena exhibited by nature" (p. 61) before presenting these data in other works for the public. He hopes that his tableau will attract a public not only of scientists but will "also stimulate people to study it who do not yet know all the pleasures associated with developing our intelligence" (pp. 79–80), and that it will be "capable of speaking to the imagination and providing the pleasure that comes from contemplating a beneficial as well as majestic nature" (p. 79). He hopes that this plate will encourage others to make similar plates not only for the temperate regions of Europe (p. 94) but also for the hotter zones of Africa and the East Indies (p. 99). Finally, just before he describes in detail the scales framing the picture, he sums up his hope that "it could also help us understand the totality of our knowledge about everything that varies with the altitudes rising above sea level" (p. 99). Publicity, the need to attract funding (which the wealthy Humboldt did not have to worry about), and especially the dissemination of knowledge go hand in hand for Humboldt, much as they do for today's scientists.

It is worth noting that this plate is Humboldt's first large-scale illustration, and that he continued afterwards to be a tireless and inventive illustrator. In his *Atlas of New Spain* that came out of this same voyage, he continued experimenting with visual illustrations, cross-sections, maps, and graphs. All these illustrations, but perhaps none more forcefully than this first one, demonstrate Humboldt's clear demand for a holistic approach to nature, the Humboldtian science uniting both measurements and distribution of the "broader phenomena exhibited by nature," to arrive at the "comprehensive view of these phenomena" (p. 61) as announced on the very first page of the "Essay." It is, as John Tresch (forthcoming) states, the first of the "synoptic tableaux that Humboldt invented" and one of the most comprehensive, a "whole system . . . a balancing act between opposed forces."

While Humboldt's plate represents a new type of scientific picture, his drive to measure and illustrate has its roots in its historical and cultural context, especially in the eighteenth century's interest in illustrating scientific and technical knowledge.

ART, SCIENCE, AND VISUALIZATION
IN THE LATE ENLIGHTENMENT

The use of visual depictions is not new to science: since the Renaissance, maps have been drawn to show newly discovered lands; plants have been illustrated in herbals; Galileo and Leonardo da Vinci were profuse sketchers and illustrators. Descartes was a pioneer in using diagrams in his scientific discourses and devised his graphs of coordinates to show abstract quantities in "terms which are unashamedly spatial and, indeed, pictorial" (Gaukroger 1995, 176). The development of printing and engraving made images easier to reproduce with precision, and from the sixteenth century on, a larger, well-educated elite could afford to buy prints and appreciated the acquisition of learning both for its own sake and for its practical aspects. The momentum gathered strength during the Enlightenment, which, as many historians have noted, witnessed a more generalized use of illustrations used for instruction. Wainer states that the "graphic explosion of the nineteenth century . . . had its origins in the intellectual turbulence of the eighteenth century: the hundred-year span between 1750 and 1850 saw a shift in the language of science from words to pictures" (Wainer 2005, 5). Humboldt was living in Paris, one of Europe's "centres of calculation," to use Bruno Latour's term (Latour 1987), where innovative thinkers were gathered, and so he was well placed to witness and absorb the new visual practices which were being developed at that time in this city.

The expanded use of graphics in second half of the eighteenth century arises from several developments in art that encouraged the creation of new forms of visual representation. One major development is an esthetic evolution that cultural historians have described as a weaning of art from an obligation to please as well as instruct, to be "dulce et utile," decorative and useful. This earlier ideal was based in an aristocratic mindset, in which nobility of purpose went with nobility of behavior, in which the arts were united in a common purpose that aristocrats could understand but which remained opaque to people not schooled in elite manners and values. This visual culture, generally called baroque and associated with the "private conceits of an aristocratic court culture," "became tarnished during the Enlightenment" (Stafford 1994, xxv). Gradually during the eighteenth century a growing, wealthy, and powerful middle class of merchants, lawyers, intellectuals, traders, and manufacturers, which increasingly also included aristocrats, did not participate in the aristocracy's traditional codes of conduct and artistic representation. As Erica Harth states, "specializations that arose with merchant capital and early modern science and technology had the effect of severing science from art" so that the aim of art became "to represent the world not as it seemed to be but as it was" (Harth 1983, 310–11). The obligation to be ornate gradually faded over the course of the century, though not completely: one small remainder can be seen in the very ornate lettering of the dedication to Jussieu and Desfontaines on the preliminary page of the volume. And as noted earlier, Humboldt organizes shapes, colors, and words in a harmonious fashion.

In addition, esthetic tastes generally evolved towards a simpler style in the later part of the century that in France is called "Louis XVI" style, characterized by simpler straight lines, less of the curving foliage and flowers favored in the earlier, more elaborate "rococo" style. In the last decades of the century this evolved into the new neoclassical style, as represented, for example, by the painter Jacques-Louis David (1748–1825) and his more austere and linear depictions of heroic Roman scenes.

The process of representing objective, unornamented truth had a subversive impact in a society where many still felt bound by the old rules and codes inherited from the seventeenth century, codes that faded over the course of the eighteenth century but never completely disappeared. A tension remained between Enlightenment culture and surviving modes of representation where pleasure and instruction still went hand in hand. As Barbara Stafford states, "the undemanding 'spectacle of nature'—encouraging an effortless gaping among the leisured classes—was challenged in the second half of the century by new professionals forging exacting taxonomies" (Stafford 1994, 218). Yet, "for eighteenth-century thinkers, the role of pleasure in the formation of an 'educated sensibility' remained troublingly equivocal" (Stafford 1994, 73). The upheavals of 1789 were radical, but they were preceded by longstanding tensions in society and they only hastened the process of change—the French Revolution forms "another phase in the development of a new society" (Harth 1983, 310) that deliberately turned its back on its past with a new set of values, such as objective truth, equality, secularism, and nationalism.

Science also underwent major developments as it shifted from deductive reasoning and Cartesian rationalism to the empiricism of Locke and other British empiricists who rejected the principle of innate ideas and emphasized knowledge gained through the senses and through induction. Sensationalist philosophy emphasized collecting large amounts of discrete sensory inputs, and the analytic approach stressed breaking down phenomena into small parts and recombining them into understandable wholes. The goal was "to understand how complex judgments and discourses were the outcome of primitive sensations through a process of gradual abstraction" (Picon 2004, 86). In France, the movement towards empiricism was widespread and diverse, and it led to some bold, materialist man-as-a-machine concepts.[10] Freeing representation from the obligation to be ornamented and valuing empirical data led on the one hand to exactitude in observation and on the other to a search for innovative ways to show the data. One should be more precise, but one could also be more experimental. The contradiction is only apparent, as the emphasis of empirical data led not only to an increasing accumulation of many tables of data but also to a desire to persuade by means of these data: at a time when a larger audience for learning existed, better means of persuasion and promotion were being devised in science as well as in literature. Another important consequence of this proliferation of data led to a debate about the integration of these data into a coherent system of knowledge. It is at this time that the modern notion of structure was being elaborated, for as Antoine Picon states, "until the eighteenth century, at least in France, a structural attitude was hard to identify" (Picon 2004, 94). He reminds us that structure has a specific "complexity and cultural character" (94), and Humboldt's holistic depiction of numerous individual data and their situation in large ensembles can be seen as a locus of reflection on structure.

The Enlighteners' drive to spread their new ideas about human nature, government, and freedom of religion and speech led them to invent new literary forms or modify old ones: plays showing average people rather than monarchs, adventure and epistolary novels, philosophical dialogues, allegorical tales, satiric poems—all these were used by the *philosophes* as they explored the ideas that led to the upheavals of the French Revolution. Similarly, they devised innovative kinds of illustrations to disseminate new

10. Empiricism was not the only current of thought in the eighteenth century, but it was a dominant one; for an exposé of the various currents of thought at that time, see Bowen's useful discussion of positivism, sensationism, mechanism, and more dynamic views, as well as forms of idealism like the Kantian one (Bowen 1981, especially chapter 6, "Science and Philosophy: Enlightenment Conflicts in Europe," 174–209). On the importance of vitalism, a scientific current between the purely mechanistic one and the later Romantic, idealist one, see especially Peter Reill 2005.

knowledge in such varied domains as economics, manufacturing, biology, medicine, geography, and geology.[11] In the second half of the eighteenth century several notable visualizations of knowledge were created, some of which Humboldt might well have seen in Paris, and some that he certainly did see.

The most formidable and comprehensive effort to represent visually almost every aspect of human life and activity was Diderot's *Encyclopédie*, which in its final form included 2,569 plates in eleven volumes published from 1760 to 1772. Illustrations were central to the *Encyclopédie* from its conception in 1750, when it was projected to include two volumes of plates in its ten volumes.[12] They depict every possible facet of contemporary life: nature, medicine, commerce, the arts, and especially manufacturing, testifying to the beginnings of the Industrial Revolution. They show in detail how to make such ordinary items as stockings, paper, jewelry, and playing cards, how to build a mining shaft or a theater, how to remove a kidney stone or build a ship—they were put on a par with the volumes containing philosophical discussions of religion, statehood, history, morals, and literature. The plates devoted to the techniques of manufacturing are elaborate illustrations depicting in minute detail each and every part of a machine or a mill or a manual process, as well as the machine in its totality and the various stages of the work of human and sometimes animal labor. They were designed to give the maximum clarity of representation through the use of profiles, scales of measures, and pictures of each individual part of a machine, thus parsing the manufacturing process in the successive stages of human motions and assembly of materials, with careful references and captions. So that anyone could easily access this information, they were organized alphabetically. Thus the mechanical arts coexisted with history, letters, and philosophy, and nature coexisted with strenuous manual labor. The result is, as Jacques Proust (1985) states, that

11. Dissemination of knowledge among a wide public became central to Enlightenment society, where a new public sphere of opinion and discussion was constituting itself. On the public sphere's growth and significance, see Habermas's central study (1989). On the *Encyclopédie* as a commercial enterprise, see especially Darnton 1979.

12. For these details regarding the plates, I am drawing on Werner 1993. On the importance of the *Encyclopédie* in reorganizing and disseminating knowledge, and on the commercial modes of such diffusion, see Darnton 1979 and 2001: "Having mapped the world of knowledge, the Encyclopedists needed to conquer it; and to do so, they had to get their book into the hands of readers" (Darnton 2001, 66).

"reason can account for the totality of what is observable; the human mind can acquire a representation of things that is totally comprehensive and at the same time completely intelligible."[13]

The *Encyclopédie*, with its plates occupying a central position, represents a new genre. As Werner (1993) states, the difference between previous illustrated compendia of knowledge and this one is the "difference between an illustrated encyclopedia and a pictorial one." Diderot sought to "break away from [an] enfeebled conception of art" and ensure that the plates might be "validated for their sake, promoted to a position of prominence in the new French encyclopedia. . . . This new genre can be most satisfactorily termed a *pictorial encyclopedia*" (117–18, Werner's emphasis). One difficulty with the *Encyclopédie*'s use of visual representations, was the relationship between the plates and the text, that is, with Tufte's criterion of integration between the verbal and the visual: Diderot "did not succeed so well in coordinating text and image" (Proust 1965, 174). This became even more problematic as the Encyclopedists' project developed, and the images that were destined to accompany the articles became more separate as the project grew enormously in size. The hoped-for holistic view of knowledge remained visible in principle but not so clear in practice. Humboldt strove for a much stronger and tighter integration of text and image, in keeping with his vision of an analytical as well as a holistic, visually oriented science.

A greater variety of more specialized graphical representations of data were being developed in the same period. Tufte (1997) says that "by 1765, two-dimensional space was liberated from pictorially-based scales" (15), and that the "two-dimensional plane" was "available for any measured data" (16). This freedom was actually already achieved earlier than this by Descartes. But in the eighteenth century, there was an explosion of many types of innovative illustrations. For example, a famous one-page illustration of economic relations was published in 1758 by François Quesnay (1694–1774), a writer of several articles for the *Encyclopédie* and a member of the physiocrat movement in economics. It appeared at a time when "political economy . . . emancipated itself from its roots in moral and political philosophy" and social science was "freed from the theological encumbrances" of earlier centuries (Groenewegen 2002, 3). Quesnay's work, entitled *Tableau économique*, is a one-page representation consisting of three columns, related by a zigzag of

---

13. Proust 1985, n.p. See this work also for detailed discussions of 38 of the plates.

dotted lines showing the relationships between "productive" spending (i.e., produced by agriculture), and nonagricultural or "barren" spending. During his career, Quesnay continued to refine his *Tableau économique* and presented four more versions of it and inspired others to imitate him. Suaudeau situates Quesnay as an inspiration to "today's virtuosos of scientific vulgarization and economic propaganda" (1958, 9). His representations of the relations between production, income, and spending are completely new in that they are not quantitative; they have a "mathematical look" while being "*qualitative* representations that we are so used to seeing in today's newspapers, magazines, tracts, posters that they have become banal" (Suaudeau 1958, 8–9).[14] One can speculate whether Humboldt had seen Quesnay's *Tableau économique*—but certainly Quesnay made the word "tableau" common among intellectual and government circles, and his graphic tableaux were widely known in that period.

Much less speculative, as far as Humboldt's knowledge is concerned, are the very innovative graphs devised by an English economist who resided in France from 1787 to 1793: William Playfair (1759–1823), who, in his *Commercial and Political Atlas* (1787) and his *Statistical Breviary* (1801), single-handedly invented many forms of visual representation of data, such as hachures, color coding, bar graphs, gridlines, pie charts, and line charts (Wainer and Spence 2005, 23–27). The example of Playfair's new ways of graphing data is particularly significant since Humboldt mentions Playfair by name in his *Analyse raisonnée* (Commentary on Plate XX) and in his *Essai politique*, chapter 1. "Humboldt thought highly of his [Playfair's] creations" (Wainer and Spence 2005, 31)—indeed Playfair was "received more kindly on the continent" (Wainer and Spence 2005, 31), especially in France where he saw his work translated into French at that time. Clearly, Humboldt is very aware that he is exploring the newest modes of graphic representation, as he correctly categorizes himself among those who have thought about graphics: "Those who have reflected about graphic methods and who have attempted to improve them, will realize, as I did, that such methods can never be completely satisfactory" (*Analyse raisonnée*, Commentary on Plate XII). But these new modes are also capable of communicating

14. Suaudeau's emphasis. He reproduces Quesnay's five versions of his *Tableau économique* and his two other interpretive schemas on the circulation of wealth (with and without the intervention of money) that use interlocking circles modeled on Harvey's circulation of the blood, as well as a triangle devised by d'Alembert.

information and ideas about such varied events as international trade and epidemics. As Wainer and Spence state in their introduction to the reprint of Playfair's works, this is the time when some longstanding "18th-century barriers to statistical charts" started to fall, and when the "inspiration to represent them as pictures" could start to be imagined (Wainer and Spence 2005, 9–10).

Lastly, the most relevant precedents are some images by Charles-Marie de La Condamine (1701–74) in two works on measuring a section of the Quito meridian: his *Journal du voyage fait par ordre du roi à l'Équateur servant d'introduction historique à la mesure des trois premiers degrés du Méridien* (Journal of the Voyage made to the Equator on the King's Command, as an Introduction to the account of the Measurement of the First Three Degrees of the Meridian) and in his *Mesure des trois premiers degrés du Méridien dans l'Hémisphère austral* (Measurement of the First Three Degrees of the Meridian in the Southern Hemisphere) (1751). These works are important here for several reasons. They concern the same terrain, and Humboldt knew La Condamine's work even though his own explorations were not the same as his predecessor's geodesy. And both contain complex fold-out charts and profiles of the mountains in Humboldt's plate. Two particular plates showing profiles in each of these two works seem important to understand Humboldt's innovative reworking of La Condamine's illustrations.[15]

Plate II of the *Journal* (fig. 8) is titled "View of the base line measured in the Yaruqui plain from Caraburu to Oyambaro, in an arc comprising 180 degrees along the horizon, drawn from the top of a waterfall of the fuller's earth mill." Measuring 43 cm by 19.5 cm, it is a horizontal rectangle divided across into two fairly equal parts. The upper part shows the mountains on the horizon as seen from the higher vantage point of the nearby waterfall. The cross-section goes from the southern point of the initial triangle's base at Caraburu to the northern end of the base at Oyambaro, indicated clearly by two little pyramids and given pride of place in the lists in the lower part of the page that contains three columns listing the names of the reference points and the most prominent objects. This panoramic view aims to be realistic in depicting three-dimensional space, with trees in the foreground, foothills in the middle distance, and mountains at the horizon. In the upper left corner,

15. For a brief history of the use of profiles in the eighteenth century and other precedents to Humboldt, see Wolter 1972, who treats only the *Atlas*'s profiles, not the *Essay*'s.

Pl. II. Introd. Part. Pag.

Vue de la Base mesurée dans la plaine d'Yarouqui, depuis Carabourou jusqu'à Oyambaro,
Sous un arc qui comprend 180 Degrez de l'horizon,
Dessinée du haut de la chute d'eau du moulin-à-foulon d'Yarouqui.

FIGURE 8. "Vue de la Base mesurée dans la plaine d'Yarouqui depuis Carabourou jusqu'à Oyambaro, sous un arc qui comprend 180 degrés de l'horizon, Dessinée du haut de la chute d'eau du moulin à foulon d'Yarouqui" [View of the base line measured in the Yaruqui plain from Caraburu to Oyambaro, in an arc comprising 180 degrees along the horizon, drawn from the top of a waterfall of the fuller's earth mill].

This is second of six plates folded into La Condamine's Journal, a work written for the general public; other plates of varying sizes scattered throughout the volume show the routes of La Condamine's journeys in 1735 and 1744; a map of the city of Quito; an allegorical representation of the Paris Academy's scientific and mathematical works; diagrams and profiles of the two pyramids elected at both ends of the first triangle's base; and a map of the province of Quito. All the illustrations except the allegorical one and the second plate have scales or indications of distance.

This figure shows a panoramic view of the Andean landscape, with an unobtrusive overlay of some scientific sites. There is no indication of scale. References to mountains, buildings, and scientific sites are given in the three columns below the landscape. From left to right they read:

References to the signals included this picture that were used for the series of triangles of the Quito meridian.

References to the objects visible in the plain seen in this picture.

References to the objects visible on the horizon.

The signals in the left-most column are references to the points used in measuring the meridian; the first two sites, indicated by capital letters, are the two pyramids placed at the southern and northern ends of one side of the base triangle; they are linked by a dotted line running through the middle of the plain. The references of the other two columns are not scientific and refer to houses, farms, churches,

and other features of the terrain. At the top left is the diagram showing a stylized eye and the half-circle representing what can be seen from one point. Save for the buildings, the picture is empty of any human presence, emphasizing the vastness of the landscape. The spectator is perhaps not aware when looking at this figure that the distance between pyramids A and B is only the base of the first of many triangles that La Condamine and his team measured. Courtesy of the Northwestern University Library, Special Collections.

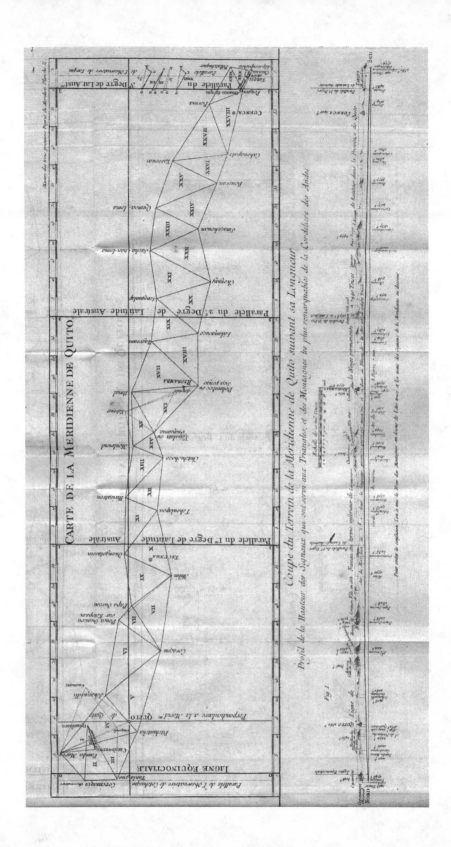

CARTE DE LA MERIDIENNE DE QUITO

Coupe du Terrein de la Méridienne de Quito suivant sa Longueur

Profil de la Hauteur des Signaux qui ont servi aux Triangles, et des Montagnes les plus remarquables de la Cordelière des Andes

FIGURE 9. "Carte de la Méridienne de Quito" [Map of the Quito Meridian] (the plate's general title) situated in the top half. The bottom half is titled: "Coupe du Terrain de la méridienne de Quito suivant sa Longueur: Profil de la Hauteur des Signaux qui ont servi aux Triangles, et des Montagnes les plus remarquables de la Cordelière des Andes" [Cross-section of the terrain of the Quito meridian, lengthwise: profile of the elevation of the signals that were used for the triangles, and of the most remarkable mountains of the Andes Cordillera].

This is the second of only three plates folded together at the back of Mesure des trois premiers degrés du Méridien dans l'Hémisphère austral, a volume focused entirely on the scientific work of the expedition. The other two plates show a schematic profile of the bases of two triangles, and an instrument, that La Condamine had built to his specifications, described in a section of his Journal that serves as a caption for that plate.

The top section of the plate shows the successive triangles used to measure three degrees of latitude from north to south, and it is neatly framed in a rectangle showing the minutes of each degree. It correlates with the bottom section showing the location of these same points on the mountains that are reduced to a thin wavy line along the bottom. Much information is crowded onto the lower part: the names of the mountains, the locations of the signals, the inferior limit of permanent snows indicated by a dotted line, and the elevation of each mountain. At the same time, clarity and ease of reading were a concern: at the bottom, there is the comment: "In order to avoid confusion, the names of the mountains have been placed above the arc, and the names of the meridian's markers below." When one looks at this diagram in conjunction with figure 8, one realizes that the vast panorama of figure 8 shows only the base of a tiny triangle at the top left of figure 9. The more realistic Journal plate is impressive enough, while the scientific plate of Mesure is perhaps more eloquent on the immensity of the geodesists' work despite its more abstract character. The lack of framing and the small scale (4.7 cm long) showing the considerable distance of 20,000 toises (38,980 meters) only emphasize the vast area covered by the scientists. Courtesy of the Northwestern University Library, Special Collections.

a diagram, with an eye and a half-circle above it, explains that this panorama could be seen by a stationary observer surveying the horizon in a half-circle. This is probably the least scientific-laden profile, aiming only to give an idea of the landscape, with no indications of scale or heights.

Much more complex and rich in scientific content is a plate titled "Map of the Quito Meridian" contained in *Mesure des trois premiers degrés du Méridien dans l'Hémisphère austral* (fig. 9). Despite its very different appearance from Humboldt's plate, I believe that it is quite relevant as a precedent. A horizontally oriented rectangle (48.5 cm by 27.3 cm), it is divided into two parts that are interrelated visually. The upper part (14.2 cm high) shows the triangulations from the equator south to the third degree. A line at each degree crosses vertically into the lower part of the diagram, a "coupe du terrain" [cross-section of the terrain] (5 cm high) showing the corresponding mountains in tiny dimensions: they look like a continuous wavy line upon first sight, and Chimborazo is no more than 6 mm high. Humboldt's remark that if his depiction were true to scale, "Chimborazo would be only 4 millimeters high" (p. 85) can suggest that he looked at this plate and sought to overcome the problems of representing mountains faced by La Condamine. Below the mountains' profiles, the names of the places where the main signals were located are correlated with the mountains' profiles and the triangles in the top part. This does not show a realistic, recognizable landscape like the *Journal*'s plate and like Humboldt's plate, which is somewhat realistic in its central panel.

Though visually La Condamine's *Mesure* plate does not resemble Humboldt's, I think that several features of its construction can be compared to it. In both plates, there is no attempt to represent three-dimensional space, and the mountain profiles are highly schematic. It is as if Humboldt looked at the correctly scaled representation of Chimborazo, zoomed in on two of La Condamine's tiny mountains, and blew them up to large proportions, just as he made his plate much larger (81 x 36.5 cm). In both plates, there is a good deal of information relating to measurements, and a juxtaposition of the scientific data to the landscape that makes the eye travel back and forth, up and down in La Condamine's and across in Humboldt's. In both plates, the structure carries a similar message: there is a holistic approach that conveys the spatial correlation of measures, places, and landscapes—lines, names, and triangles are imposed on profiles of mountains and have to be understood together. Thus it is in the intent and the message about their authors' view of scientific exploration and activity that these two plates, though dissimilar in appearance, resemble each other regarding their underlying, overall structure and meaning. If La Condamine was a Humboldtian scientist *avant la lettre* half a century earlier, it could also be said that Humboldt stood on his shoulders as he explored, measured, and represented the same equatorial landscapes.

Thus the *Tableau physique* draws on many aspects of pictorial work carried out in the eighteenth century. The *Encyclopédie* provided numerous volumes of exact and detailed plates that stood alone, apart from the articles, and imparted types of knowledge that the articles could not. Quesnay's *Tableau économique* was a striking example of an abstract representation of an entire economic system on one page by means of lines and words. Playfair was the first to devise new graphic forms to represent economic and political relationships. Other examples Humboldt would have seen, given his education and his work in mining, would include drawings of mines and geological formations.[16] Finally, there is what I believe to be one of his main inspirations, La Condamine's profiles of the equatorial mountains which he drew on and transformed radically.

To sum up what we have seen so far: the plate and the text are two interrelated tableaux, and the plate is a structure whose differentiated parts stand alongside each other while encouraging the spectator to look at them together. Science had moved away from Cartesian, deductive systematicity and instead placed primary value on gathering facts and on sense-based empiricism. Art gradually became freer from an obligation to be ornate and pleasing to an aristocratic elite, and strove both to pay close attention to the depiction of the world and to be accessible to a wider, educated public—this favored the development of many forms of visual representations and new types of illustrations and graphs. All these historical developments indicate an approach to science and to its visual representations that is both detailed and global, analytic and synthetic, empirical and holistic. One of the major questions in the late Enlightenment was that the large quantity of newly discovered phenomena and scientific data "threatened chaos," and as a result, "diversity had to be pulled into order" (Porter 1980, 309).[17] This essay will continue with a brief survey of holism and the debates surrounding it by situating it in its philosophical context.

16. For some examples of such illustrations and a discussion of them, see Rudwick 1996 and Stafford 1984, where one can see numerous reproductions of landscapes, mountains, and profiles from the eighteenth and nineteenth centuries.

17. Porter's study (1980) is a valuable and well-documented overview of the history of eighteenth-century environmental science, its relative neglect till 1980, and suggestions for future scholarship.

HOLISM: THE SEARCH FOR THE RIGHT ONE

It is obvious that Humboldt was from the beginning of his career strongly in favor of a holistic approach to nature, as evidenced by this plate, his statements in the "Essay," and the well-organized relation between the texts and the plate. Sometimes his holistic approach is attributed to the influence of the Romantic movement represented in literature by Schiller and Goethe and the movement in the history of science and ideas that derived from the Romantics known as Naturphilosophie.[18] It lies beyond the scope of this essay to address this question fully. However, Humboldt's position and strong call for a holistic view of the study of nature can also be viewed as arising from concerns emerging within the late Enlightenment that led eventually to Romantic views.

Throughout the second half of the century, the Enlightenment philosophers discussed the need for a unifying system for the sciences, as well as for all branches of human knowledge. There were competing views in every science on how to organize the increasingly large numbers of observed facts that were being gathered and produced by the continued exploration of the earth and the development of ever more precise instruments. Earlier in the century and in the previous century, classification was a powerful provider of overview and order, as it had been, for example, for La Condamine, who defined botany as "the work required to describe these plants exactly, and to perform their grouping [leur réduction] in classes, genera and species" (La Condamine 2004, 72). At the time of the Encyclopédie, taxonomy, widely accepted as necessary even if still debated, no longer seemed sufficient. Earlier, measuring and naming had seemed like progress, but towards the end of the century what had been progressive came to be considered narrow, divisive, and reductive: witness Humboldt's opening statement about botanists that reads like an indictment: "Botanists usually direct their research towards objects that encompass only a very small part of their science. They are concerned almost exclusively with the discovery of new species of plants, the study of their external structure, their distinguishing characteristics, and the analogies that group them together into classes and families" (p. 64). The goal of taxonomy was clear, though the criteria of how to classify were not entirely settled yet in the middle of the century, when Linnaeus was challenged by such eminent scientists as Buffon. The competition between taxonomies encouraged thinkers to search for more general systems of knowl-

---

18. For a good summary of Naturphilosophie at it relates to the sciences, see Jardine in Jardine, Secord, and Spary 1996.

edge, as did Humboldt, who sought to go beyond problems of classification to the building of a "Geography of Plants, a science that up to now exists in name only" (p. 64).

The French Enlightenment thinkers debated the merits of building systems of knowledge, and at the same time they worried about building the wrong kind of systems. Systems are "fraught with danger" by imposing an order that is too rational to take into account the heterogeneity of nature, and yet they "have produced undeniable results" (Hayes 1999, 54)[19] by stimulating research and by furnishing practical tools for understanding nature. Humboldt certainly sought to grasp the cosmos systematically. I would like to discuss first one important work concerned with the organization of scientific knowledge into systems, then, more briefly, other critical voices of the period that had an influence on Humboldt.

In the preface to the *Encyclopédie*, the *Discours préliminaire* (Preliminary Discourse) ([1751] 1995), its author, Jean Le Rond d'Alembert (1717–83), outlines "two aims": "to set forth as well as possible the order and connection of the parts of human knowledge," and to "contain the general principles that set forth the basis of each science and each art, liberal or mechanical, and the most essential facts that make up the body and substance of each" (1995, 4). He outlines two ways to organize this content: one is to trace "in historical order the progress of the mind," and the other is to provide an "encyclopedic arrangement" "into the smallest area possible": the *Encyclopédie* will provide "a kind of world map which is to show the principal countries, their position and their mutual dependence" and show the links between these "countries" by means of "individual, highly detailed maps. These individual maps will be the different articles of the *Encyclopedia* and the Tree or Systematic Chart will be its world map" (47–48). The "individual maps" are the texts of the *Encyclopédie*; the "systematic chart" is the large synoptic tableau, the "Système figuré des connaissances humaines" [Detailed System of Human Knowledge] that is included in the *Discours préliminaire*, showing the affiliations of the various fields of knowledge by grouping them under the three broad headings of Memory, Reason, and Imagination (reproduced in d'Alembert 1995, 144–45). There is a tension between the massive volumes (texts and plates) and a mere two-page Tree of Knowledge that constitutes a synoptic view of that knowledge organized in one work, albeit a rather huge one. As Julie Hayes states, there is an underlying "fissure between discursive

---

19. For a discussion of these concerns, see Hayes 1999, 22–57, especially chapter 1, "'Système'—origins and itineraries."

succession and figural simultaneity," between "incommensurable elements" of diametrically opposed size and scale in the system (Hayes 1999, 39).

D'Alembert's approach to mitigating the tension between the needs for rigor and flexibility is to allow each science to develop separately. The "individual maps" (d'Alembert 1995, 47), that is, the individual sciences, are to be "highly detailed," for he is wary of "hypothesis and conjecture" and wants to confine sciences "solely to observations and to calculations," for example, to confine "natural history to the detailed description of vegetables, animals, and minerals" (94–95). The systematic unity that the *Encyclopédie* will provide is only at the level of the relations among the sciences, not within the individual sciences themselves. Thus there is a division of labor: scientists measure and calculate, and the philosopher who enjoys "a vantage point, so to speak, high above this vast labyrinth, whence he can perceive the principal sciences and the arts simultaneously" (47) can see the totality of human work. The scientists study the details, whereas the philosophers see the big picture.

In his explanations of the "Système figuré," d'Alembert provides many clear examples of this program for how each science is to proceed and advance. Botany is situated under the heading "Science of Nature" and the subheading "Particular Physics," and its progress is described thus: "From the history of *plants*, perceived by the senses, reflection has gone on to investigate their economy, propagation, culture, vegetation, etc., and has engendered *botany*, of which *agriculture* and *gardening* are two branches" (154–55, his emphasis). Although the directions in which botany should go according to d'Alembert, agriculture and gardening, were moving beyond classification, they could not have seemed much more than mere classification to someone like Humboldt. What Humboldt imagines for a specific science of plants is precisely what the *Encyclopédie* proposes at the general level of human knowledge. Humboldt's new science will be synthetic, like the synoptic tableau of "Human Knowledge" presented in the *Discours préliminaire*, and in Humboldt's vision of science, a "particular physics" becomes "general physics." There will be no division of labor, and scientists will also be philosophers.

In the latter part of the century, however, there was both an increasing specialization of science and a desire for a universal science. As Roy Porter states, "New restricted sub-disciplines were grouping toward the end of the century . . . yet hopes remained of creating a new universal science of the environment" (Porter 1980, 320–21). Some scathing critiques of specialization were made by such figures as Immanuel Kant (1724–1804) and Friedrich Schiller (1759–1805), who voiced disappointment with what seemed to them an overly narrow practice of science. For example, Kant, who lectured for many years on geography and who influenced Humboldt (May 1970, 75–78): "There is a great lack of instruction as to how one may use the previously acquired knowledge. . . . Furthermore we have to know the objects of our experience as a whole so that our knowledge does not form an aggregate but rather a system; in a system it is the whole that comes before the parts, whereas in an aggregate the parts are first" (Kant in May 1970, 257). In an early work, On the Aesthetic Education of Man in a Series of Letters ([1795] 1954), Schiller bemoans the "culture that inflicted this wound upon modern humanity," meaning a wound that was inflicted upon "all-uniting Nature" by an "all-dividing intellect"; speaking about various philosophers of various intellectual persuasions, he laments that there is too much specialization, "a sharper division of the sciences," and "hostile attitudes upon their [the scientists] respective fields, whose boundaries they now began to guard with jealousy and distrust" (Schiller [1795] 1954, Sixth Letter, 39).[20] Schiller's

20. My quoting of Kant's and especially Schiller's critiques of their age does not mean that I propose to see Humboldt primarily as a Romantic, much less as a follower of the Naturphilosophie movement. This topic lies beyond the purpose of this essay. Opinions are very divided among critics, some putting Humboldt in the Romantic camp (for example Richards, Tresch), some asserting his belonging to the Enlightenment and minimizing his Romantic affinities (e.g. Cannon, Minguet), for example: "One principle prevails: one must measure, weigh, calculate the phenomena of nature . . . Humboldt studies the myriad aspects of nature in America, remaining faithful to the rational empiricism that he constantly referenced" (Minguet 1997, 377). Many see a balance between both types of thought (e.g., Bowen, Castrillon, Minguet, Nicolson, Reill, and Wise and Wise): "Humboldt, although always alive to the prerogatives of aesthetics and the appeal of the sublime, did not follow Schiller or Schelling in subordinating rationality to aesthetic sensibility. . . . [T]o Humboldt, aesthetics complemented rationality" (Nicolson 1990, 180). For a judicious and complete assessment of all the attributes and qualities given to Humboldt over the past two centuries, see Rupke 2005. See also Sachs (2003, 124), who sees Humboldt as an Enlightenment man of science in whom concern for exact measurement was balanced by, and co-existed with, Romantic idealism and sentiment throughout his entire work. In the Introduction to Cosmos, Humboldt was to state his commitment to base his study on a "rational empiricism, that is to say, upon the results of the facts registered by science" (49). The "unity" of the "great phenomena of nature" is not "deduced from ideas alone" (49–50). He inveighs against "the mythical idea . . . of the imponderable substances and vital forces" (75) and some current tendencies that lead "our noble but ill-judging youth into the saturnalia of a purely ideal science of nature" (76). He reaffirms forcefully that in his scientific career he has been engaged "in measurements, experiments, and the investigation of facts" and that he limits himself to "the domain of empirical ideas" (75).

Letter illustrates the tension between the accumulation of "more precise" knowledge and the desire to "restore by means of a higher Art this wholeness in our nature which Art has destroyed" (Schiller, Sixth Letter, 39, 45).

Finding the right universal science proved arduous: "One sees the formulation of heroic strategies, and the building of ambitious—precarious—structures" of which Humboldt's "ideal of articulating the harmony of the cosmos, unity in diversity" was one (Porter 1980, 322). During this period, the conception and the practice of knowledge (alongside the political upheavals of the time in France and Europe) were changing "to focus more clearly on forces and origins" (Browne 1983, 41). Michel Foucault (1966) has characterized this shift as the passage from an age of classification (often labeled the age of representation) to what he calls the modern age. In the latter part of the eighteenth century, classification, while still useful, seemed to have reached its limits, degenerating into competition among systems of classification. Gradually another conception emerged, based on an interior principle that was dynamic, and that principle takes into account hierarchical organization, and the capacity to adapt and change over time. In the domains of language, labor, and life, "the general area of knowledge is no longer that of identities and differences, . . . that of a general *taxinomia* . . . but an area made up of organic structures, that is, of internal relations between elements whose totality performs a function" (Foucault 1970, 218). This interior organization is not visible to the eye, as were the characteristics that could identify organisms, but requires a new understanding of nature as dynamic and continuously self-transforming, in order to seize "the coherent totality of an organic structure that weaves back into the unique fabric of its sovereignty both the visible and the invisible" (229). Classification depended on identifying similarities and differences that were visible on the surface; the internal organization is invisible, depending on the "functions, which are not perceptible, but determine, as though from below, the arrangement of what we do perceive" (268). This entails a shift in scientific work, which Foucault describes in terms that can apply to Humboldt's own undertaking in the Andes: "In order to discover the fundamental groups into which natural beings can be divided, it has become necessary to explore in depth the space that lies between their superficial organs and their most concealed ones, and between these latter and the broad functions that they perform" (230). What Foucault states about studying living organisms as a structurally coherent whole, Humboldt imagines at the larger level of what we would call today entire ecosystems.

If the new understanding of organization of plant life, its distribution according to zones and latitudes and other factors, was based on something less visible than plant structures, why does Humboldt want to convey it in a figurative tableau? To answer this question, one can consider how he aims to move beyond the previous age of classification. This is clearly stated in the "Essay" and in the "Physical Tableau": this is a "philosophical point of view" (p. 64) that is opposed to classification, a contemplation of nature rather than its description (p. 73), which embraces nature as a totality in a "tableau," as he says in the only use of the word in this preliminary "Essay." And in his "Physical Tableau," all the words opposed to classification and description come together in one sentence: "I thought that if my tableau were capable of suggesting unexpected analogies . . . it would also be capable of speaking to the imagination and providing the pleasure that comes from contemplating . . . nature" (p. 79). In the words of Hans Vaihinger, who is drawing on the work of Kant, this is an important development: "the realization that imagination also plays a great part in science is one of the main advances of modern epistemology." He goes on to state that "the realization of this fact has recently passed from philosophy to other sciences, although admittedly never on the same scale" as in philosophy (Vaihinger 1935, 55). One can see the use of imagination in Humboldt's bold staking out of a new field of science and in his devising a new way of representing it visually.

Certainly Humboldt's hope was to elaborate a universally comprehensive organizing system, but, as Porter puts it, in the search for "a new universal science of the environment . . . no new intellectual framework succeeded (till perhaps evolutionism) in weaving together the sub-disciplines" (Porter 1980, 321). Humboldt's significance is to be found elsewhere, not so much in creating a new universal science (despite his ambition to do so), but in his method: in his emphasis on science as a complex and active work to be carried out on the terrain, to be represented in a variety of imaginative prose and pictorial tableaux, and to be shared with a large audience. In this manner, Humboldt was both heeding the Romantics' call for a unified approach to nature and remaining faithful to quantitative methods: "Humboldt brought to the romantic ideal a thorough commitment to the latest analytic techniques of French chemistry and physics and to the precision instruments on which that analysis depended" (Wise and Wise 2004, 113). As these authors point out, this joining together of Romantic idealism and Enlightenment precision was by no means so obvious or easy: "Humboldt's unification of measurement with beauty . . . was not easily purchased in Jena" (113), nor probably in some other places. That Humboldt himself, like others at that time, was asking the same question that Darwin would later is indicated by a brief sentence in the "Essay": "This same science can examine whether among the immense variety of plant forms one can recognize some primitive forms, and whether the diversity of species can be considered to be an effect of the degeneration that over time transformed accidental varieties into permanent ones" (pp. 67–68). In fact, as is well known, he stimulated Darwin, who took Humboldt's narratives with him on his voyage to South America, to explore the world in search of understanding of nature. Darwin succeeded in finding an all-encompassing framework through evolution, eclipsing Humboldt in the memories of later generations,[21] but he had huge shoulders to stand on.

HUMBOLDT'S PICTORIAL SCIENCE

Let us return to the tableau and the specific work it accomplishes. What can a picture say and do that a text cannot? In the words of basic literary criticism, it shows, whereas the text tells. Humboldt's picture shows the mountains he and Bonpland saw, and it is filled with words and sentences referring to what they climbed on and measured, the plants they encountered, gathered, collected, and catalogued. In showing and telling what was seen and understood, the plate (columns and picture) represents not only what they *saw* but what they *did*. Thus one can say that it provides a visual analogy of their complex labor in the Andes. Humboldt himself thought, as I noted earlier, that this plate would have the particular power of "suggesting unexpected analogies to those who will study its details" (p. 79).

It is worth pausing briefly on the notion of analogy. As Barbara Stafford has suggested in her work on the intersection of science and art, this concept may seem very outmoded but it is worthy of reconsideration. It is a mode of cognition that is visual, combinatory, and utilizes resemblances: "analogy is the vision of ordered relationships articulated as similarity-in-difference" (Stafford 1999, 9). Foucault situates resemblance as being characteristic of the Renaissance worldview ("episteme" in his vocabulary), the age that preceded the age of classification, which itself preceded the Enlightenment. In

21. As Aaron Sachs says, "Humboldt's diminished reputation is due in large part to the ascension of Darwinism" and the "supplanting of Humboldt's vision of a unified, harmonious world" by "Darwin's 'struggle for existence'" (Sachs 2003, 116), a harmonious world well exemplified by the plate.

the Renaissance, correspondences between the human and the cosmos, the microcosm and the macrocosm, the human body and the body politic were universally accepted as real and true—a worldview that was superseded by mechanism during the seventeenth century. Unlike Foucault, who emphasizes shifts between successive epistemes, Stafford believes that analogy has never completely gone away, and she wants "recuperate analogy" (1999, 8) for us today. The modern age is an heir of the Enlightenment, Romanticism, modernism, and postmodernism, and it prizes singularity, difference, and uniqueness, but analogy has not ceased to be a mode of cognition. Stafford suggests that, in our contemporary world of electronic networking, of global interconnectedness and ecological concerns, it is a mode of knowledge that is perhaps even more relevant: "The dense visual-verbal synthesis going on in cyberspace, and embodied in the popular clip art of the Web, has hastened the collapse of discrete media" (Stafford 1999, 182). The difference is that analogy no longer describes the state of the world as it did for the Elizabethans; having no ontological value, it can function freely in the culture as needed. I have already mentioned the strong interest in pictorial representation during the latter part of the Enlightenment. It is perhaps not a coincidence that, as the Enlightenment's empiricism and rationalism became the norm, visual illustrations—such as profiles, graphs on heights of mountains and lengths of rivers, and wall charts—were increasingly used in books and in the classroom both to illustrate the empirical gathering of facts and to counterbalance the abstractness of rationalist and intellectual modes of knowledge.[22]

Humboldt's plate exemplifies this kind of analogical mode of work and thought. In this way, it might belong to the category of "things that talk," as Lorraine Daston, who also studies the intersection of art and science, has called some "composites of different species" that "straddle boundar-

22. On the profusion of sketches and paintings of nature in all shapes during this period, see Stafford's entire body of work, and for examples, especially 1984; on the fortune of wall charts, see Bucchi 1998; on profiles and lengths of rivers, see Wolter 1972. This seems to disagree with Stafford 1994 whose subtitle *The Eclipse of Visual Education* suggests that visual learning became suspect, liable to fraud and "tarnished in the Enlightenment" (xxv). However, Stafford indicates that the Enlightenment distrusted a showy type of illustration done in the manner of the ornate art of the previous seventeenth century, the "oral-visual culture of the late baroque" (Stafford 1994, xxv). In the Enlightenment many new forms of graphics were invented, such as those of Fourcroy, Playfair, and Quesnay.

ies between . . . art and nature, person and things, objective and subjective" (Daston 2004, 21). Analogy operates in several ways, both in the work of scientific thought itself and in its public diffusion.

Analogies have an important role to play in science, in the inductive reasoning which builds models on known characteristics. Mary Hesse has argued for the particular value of analogies because they depend on "*observable* similarities between corresponding terms" (Hesse 1966, 69). Humboldt's work in the Andes illustrates very well the idea that "analogical argument is necessary only in situations where it has not been possible to observe . . . a large number of instances" (76), as the Andes gave him a rich model of nature on a grand scale: "near the equator plant shapes are more majestic and more imposing," flowers are "more beautiful, bigger, and more perfumed" (p. 74), and if there is less of the subtle "sweet feeling of springtime" than there is in Europe, there is in compensation "a spectacle as varied as the azure vault of the heavens" (p. 75). Analogies derived from his observations in the Andes, in "the grand and strongly characterized features of tropical scenery" (*Cosmos*, 348), in nature writ large, are especially useful in fields where one can compare similarities in difference, and such analogies can be crucial "in finding or inventing a kinship between . . . difficult-to-thread-together phenomena" (Stafford 1999, 171). This is what Hesse calls, more precisely, a homology, where there are "similarities of structure" (Hesse 1966, 82) that would lead "to the assumption that natural phenomena exhibit various forms of an underlying single reality" (127). Again and again, Humboldt looks for "analogies in shape that link the alpine plants of the Andes with those of the high peaks of the Pyrenees" (p. 73), and in so doing, he imagines similar proportions or relationships in places where nature is not writ so large. As he reflects succinctly in *Cosmos*, "This portion of the surface of the globe affords in the smallest space the greatest possible variety of impressions from the contemplation of nature" (33).

Such a type of thinking was useful to Humboldt in pursuing his fundamental view of nature as a whole, and it is exemplified in his development of isolines (see fig. 5), and in his insight on the distribution of plants according to latitude and altitude. This latter idea is expressed visually in the famous plate *Geographiae plantarum lineamenta* (see fig. 2), where three mountains of different sizes from the Andes, the Alps, and Lapland are shown side by side with their proportionally scaled bands of similar flora.[23]

23. See *Cosmos* in Rupke's edition (1997, xvii), for a reproduction of a German ver-

As these two latter examples indicate, Humboldt uses analogical thinking in his major graphic works. He devises the *Tableau physique* in such a way that it represents not nature, not a landscape, but the various labors and perspectives of the scientists—as geodesists, botanists, metrologists, geologists, even anthropologists. It is a composite, something like such artifacts as a collage, a well-constructed collection of elements that offers insight "into the combinatorial immensity of thinkable structures" (Stafford 1999, 171). This plate testifies to the communal way in which science is carried out. It combines Humboldt's work with that of Bonpland and his assistants, the makers of the instruments used, not to mention that of all the artists, engravers, and publishers involved: "Humboldt's knowledge of graphic techniques and his close supervision of the approximately 50 artists who transcribed field sketches ensured both the scientific accuracy and the aesthetic appeal of the resulting work" (Stafford 1984, 93). Although no mention is made of these fifty artists, the plate does hint at their behind-the-scenes work through the identification at the bottom of the page of the people who did the lettering and the engraving, and Humboldt mentions a few others in his text.

Finally, this plate uses analogy in a manner different from any other plate that Humboldt produced. Such a figure enables the mind to play freely back and forth over phenomena in a fluid manner and to juxtapose them in creative ways. In this way, the plate becomes a "loquacious" object, to use Daston's word, that gives "rise to an astonishing amount of talk" (Daston 2004, 11). The multiple elements are contained and organized by the scientific rigor of the numerous measurements and observations that must be gathered in person on the terrain: "After Humboldt, it became imperative for any serious seeker traveling to Chile or Bogotá to tackle the great Cordillera" (Stafford 1984, 93).[24]

The gathering together of many data on one plate can also be used to

---

sion of this plate and Browne (1983, 45) for the Latin version. A reproduction follows p. 46 in this volume. On the use of analogy by another contemporary scientist, Charles Lyell, see Rudwick 1978. Rudwick points out that analogies, such as those used by Lyell, are especially useful when a field is still new, "not yet strongly insulated and institutionalised" (67), corresponding thus to Schiller's wish to unify rather than divide the sciences. As Hesse points out, the kind of theoretical model used depends on the "various stages of the development of science" and the "particular scientific contexts" (1966, 129).

24. Photos of the instruments, the plants collected, the scenery, and many other details relating to Humboldt's and Bonpland's voyage, as well as articles on these topics,

make a geopolitical point. Earlier I noted the juxtaposition of the heights of European mountains, such as Vesuvius, with those of the Andes, which makes the Old World peaks look less impressive once one has seen the Andes.[25] Playfair, especially, inspired Humboldt to make such comparisons with a political point that went so far as to critique the Spanish colonial structure of the lands he explored. This is made quite explicit in the Atlas (1812) that he wrote as a result of this same voyage. Humboldt specifically refers to Playfair's illustration showing the relationship between the land surfaces of the mother country and of its colonies: "The inequality of the territorial distribution of New Spain was made evident by representing the governing authorities with squares inscribed with each other. This graphic method is very analogous to that first used by Mr. Playfair but in a very inferior manner" (Analyse raisonnée, Commentary on Plate XX). Playfair showed this by means of a superposition of circles of different sizes in several charts of the Statistical Breviary. For a use of squares (rather than circles) inscribed within each other, as Humboldt does in his twentieth and last plate in the Atlas, there is the much closer precedent of Charles de Fourcroy's Tableau poléométrique (1782), showing the proportions of urban surface areas ranging from "very small cities" (such as Mézières and Saint-Malo) to "very large cities" (London, Saint Petersburg, Paris) by means of superimposed squares going from very small ones at the top left to ever larger squares.[26]

Humboldt's use of superimposed squares to show the land surfaces of the mother country and its colonies is objective and anything but dispassionate: "This plate alone can lead those who are charged with watch over the prosperity and the tranquility of the colonies to some important considerations. These tableaux, similar to Mr. Playfair's, have something frightening about them" (Essai politique, chap. 1, p. 5). Humboldt, again using an objective rep-

---

can be found in Histoire, Sciences et Techniques (2003), the catalogue of an exhibition at the Musée des arts et métiers (Paris).

25. On the importance of the Pacific ocean and associated lands for the eighteenth-century imagination, see Frost in Pagden 2000 and Pagden's "Introduction" (2000).

26. For a small-scale reproduction, and a discussion of this chart, see Palsky 1996, 51, as well as his general discussion of the history of graphic illustration, including Playfair and Humboldt's approval of the latter, 51–58. For Humboldt's Plate XX, see Godlewska 1999a, 258. "Poléométrique" seems to be a neologism meaning the measurement of cities.

resentation, is quite aware of the power of such representations to make a political point in both a factual and a polemical manner.[27] And already in the *Tableau physique* under consideration here, the comparison between the peaks of the two continents suggests a new perspective to be gained by looking at nature in a place that gives a truer sense of its immensity that what can ever be gained in Europe. It is one more aspect by which Humboldt showed his awareness of the central power of graphic representations to make political and scientific statements through the use of visual forms that intrigue the viewers and enable the viewers to be involved and come to their own conclusions.

As I suggested earlier, this plate makes the spectator's eye circulate from side to side, from column to column, and from columns to picture. It is as if Humboldt wants to make us do some work on the page just as he did on the terrain: the data are there, side by side, needing to be read, reflected upon, and gathered together on the page and in our minds. The plate presents large amounts of materials and structures them, but it also leaves room for interpretation, and the lack of integration for which it could be criticized now becomes a strength. Similarly, La Condamine's plate in *Mesures*, with its elaborate triangulations coordinated with the schematic landscape below, made the spectator much more active in viewing it than the panoramic view of the mountains in the *Journal* plate. Seeing, working, gathering, and processing information over widely separate yet similar areas bring together many data and are the hallmark of a holistic and ecological perspective.

There is one detail, mentioned earlier, in La Condamine's *Journal* plate (fig. 8) that Humboldt modified, which shows his astuteness in using visual representation to involve the spectator and to make the spectator work while facing the page like the scientist facing the mountain ranges. In the *Journal* plate, there is a half circle with this caption underneath: "N.B. On a représenté dans cette vue tous les objets compris dans le demi

---

27. For a judicious, complete, and well-documented analysis that views Humboldt as anticolonialist as well as abolitionist, see Sachs 2003, who convincingly shows that Mary Louise Pratt's analysis of Humboldt is reductive and not admitting of nuances and exceptions when she states, for example, that Humboldt "never once stepped beyond the boundaries of the Spanish colonial infrastructure" and was steeped in the "history of Euroimperialism" (Pratt 1992, 127); Pratt fails to take into account his critique of Spanish colonial power and exploitation of resources, or his "undeniable environmentalist perspective" (Sachs 2003, 131).

tour de l'horizon en supposant que l'œil se tournait successivement vers chacun d'eux sans sortir du même point" [One represented in this picture all the objects that are present on the horizon's half circle, supposing that the eye would turn in succession towards each object without moving from its place]. La Condamine imagines that the spectator is stationary, but if the spectator is to embrace a full 180°, he must of course move his head. E. H. Gombrich has discussed the difference between moving the eye versus moving the head: "What happens if we merely move the eye but not the head?" (Gombrich 1980, 197), a difference that was being worked out and reflected on precisely during the seventeenth and eighteenth centuries. A turning head enables more information to be seen, whereas a stationary head emphasizes the "eye-witness principle . . . that no more should be shown than can be seen from one spot during one moment of time" (208). A stationary head, such as we have in front of Humboldt's picture, means we are seeing the scene from one perspective, ours, with some elements more clearly seen than others. In short, a picture presuming a stationary head situates the spectator more precisely, producing "a gain in the evocation of the experience of the imaginary eye-witness" and enhancing "our feeling of participation" (202). This suggests that Humboldt wanted to involve the spectator as a viewer of a painting, as much as he wanted to convey vast amounts of information, valuing the subjective imagination as well as the intellectual approach to the new science.

The text makes us read, and we are passive; the plate makes us work, and we become active—we do not become scientists literally, but we do become active viewers of an interesting landscape. This type of engagement enables the viewer to be active and more scientifically literate, as Jean Trumbo states: "There are two components to visual learning: the process of gaining awareness of the meaning of visuals and the process an individual uses to interpret meaning" (Trumbo 1996, 275). On the part of the scientist, it implies a "dual responsibility of visual representation as a teaching/learning tool and as an integral component in the cognitive process" (275) and this means that the scientists must think, more than they sometimes do, about "the challenge of achieving visual literacy among scientists, communicators, and the public" (280). Earlier scientists were strikingly productive and original visual thinkers, Leonardo da Vinci being a prime example, but his *Notebooks* were his own private works. Descartes thought conceiving of a new science was better done by one individual than by many—though at the end of the *Discourse on Method*, he appeals to the public, perhaps one of the first scientists to reach out to a wider audience, but warily and principally for funding and confirmation of his theories. Humboldt lived in an age which valued dissemination of knowledge in the public sphere, when science was carried out collectively by large groups of researchers and actively courted an audience that had to be educated in order to be persuaded. As Bruno Latour (1987) has shown in detail, this implies a view of "science in action" as a complex intellectual and social activity with the involvement of many people, scientific institutions, governments, laboratories, and consumers.

This enlarges our understanding of Humboldt's pictorial science. Humboldtian science has been defined (Cannon 1978, 104) as consisting of accuracy of instruments and observations, a willingness to entertain new ideas, new conceptual, especially visual tools, and the application of all these to physical realities. More recently, Rupke (2001, 94) called Humboldt's development of a visual language his "spatial turn," and his movement away from static taxonomy to one of "spatial ordering of scientific knowledge" (99).[28] Anthony Pagden calls Humboldt the "self-conscious heir to . . . the Baconian project" (Pagden 1993, 84), which was to create "a true 'natural' history of mankind" (86). Taking into account Humboldt's later works, Pagden describes Humboldt's encyclopedic ambitions as looking back, "perhaps with unwarranted confidence over the long history of the attempts to describe America and through description to make sense of her" (86). However, this plate represents an earlier stage of his work: it does not impose a point of view but juxtaposes data from Europe as well as America. It is neither imperialistic or Eurocentric—if anything it is the opposite, relegating Europe to the margins of the picture—nor is it resigned to incommensurability. It proposes, rather than imposes, points of comparison, rather than aiming at "absolute truthfulness" (Pagden 1993, 87).

Certainly, Humboldt was to gain during his long life great confidence in his ability to understand, and state his views about, the totality of nature. In

28. The words "spatial turn" echo an important phrase in current philosophy: Richard Rorty (1967) succinctly states that modern philosophy has taken a "linguistic turn," emphasizing that all knowledge takes place within language and that explorations of universal truths cannot take place outside language. But, as Rupke points out, Humboldtian spatial considerations in medical geography had their limitations, and "faded long before the germ theory of diseases led to a loss of interest in physical-environmental causes" because they "focus on climatological causes" of illness and overlook social causes that need "medical reform" (Rupke 1996, 310). The spatial turn was and remains important, witness the common use of isobars and isolines devised by Humboldt, but it could not provide a new overall framework for knowledge.

this early work, as a result of his voyage in the equatorial regions, Humboldt is already seeing a larger dimension of nature, and he is also aware of the challenge this dimension poses to current scientific thinking and representation, and hence is willing to be experimental and questioning. The words of Vaihinger, the philosopher of "as if," are fitting here: this plate is "not a *picture* of the actual world but an *instrument* for grasping and subjectively understanding that world" (Vaihinger 1935, 63, his emphasis). In the words of Calvino, Humboldt was able to "concentrate the part of me that was outside (and even the interior part of me that conditioned the exterior) to give rise to an image" (1968, 150). The pictorial science of Humboldt's plate exemplifies on a single page the precision, the comparative and encyclopedic nature of this work that call upon the subjective, creative, and imaginative capacities of both the scientist and the spectator. It is a picture *of* his science as well as a picture *for* science as an interactive, collaborative, and open-ended enterprise.

REFERENCES

Alder, Ken. 2002. *The Measure of All Things: The Seven-Year Odyssey and Hidden Error That Transformed the World*. New York: Free Press.

Bowen, Margarita. 1981. *Empiricism and Geographical Thought: From Francis Bacon to Alexander von Humboldt*. Cambridge: Cambridge University Press.

Browne, Janet. 1983. *The Secular Ark: Studies in the History of Biogeography*. New Haven: Yale University Press.

Bucchi, Massimiano. 1996. Images of Science in the Classroom: Wall Charts and Science Education, 1850–1920. In *Visual Cultures of Science*, 90–119. See Pauwels 2006.

Calvino, Italo. 1968 (1965). *Cosmicomics*. Translated by William Weaver. New York: Harcourt Brace Jovanovich.

Cannon, Susan Faye. 1978. *Science in Culture: The Early Victorian Period*. New York: Dawson and Science History Publications.

D'Alembert, Jean Le Rond. [1751] 1995. *Preliminary Discourse to the Encyclopedia of Diderot*. Translated and with an introduction by Richard N. Schwab with Walter E. Rex. Chicago: University of Chicago Press.

Darnton, Robert. 1979. *The Business of Enlightenment: A Publishing History of the Encyclopédie 1775–1800*. Cambridge, Massachusetts: The Belknap Press, Harvard University Press.

———. 2001. Epistemological Angst: From Encyclopedism to Advertising. In *The Structure of Knowledge*, 53–75. See Frängsmyr 2001.

Daston, Lorraine, ed. 2004. *Things that Talk: Object Lessons from Art and Science*. New York: Zone Books. Introduction: Speechless, 9–26.

Dettelbach, Michael. 1996a. Global Physics and Aesthetic Empire: Humboldt's Physical Portrait of the Tropics. In *Visions of Empire*, 258–92. See Miller 1996.

———. 1996b. Humboldtian Science. In *Cultures of Natural History*, 287–304. See Jardine 1996.

Foucault, Michel. [1966] 1970. *The Order of Things: An Archaeology of the Human Sciences*. New York: Pantheon Books. Originally published as *Les mots et les choses*. Paris: Gallimard, 1966.

Frängsmyr, Tore, ed. 2001. *The Structure of Knowledge: Classifications of Science and Learning Since the Renaissance*. Berkeley: University of California, Office for History of Science and Technology.

Frost, Alan. [1976] 2000. The Pacific Ocean: The Eighteenth Century's "New World." In *Facing Each Other*, 2:591–634. See Pagden 2000.

Gaukroger, Stephen. 1995. *Descartes: An Intellectual Biography*. Oxford: Oxford University Press.

Godlewska, Anne Marie Claire. 1999a. From Enlightenment Vision to Modern Science: Humboldt's Visual Thinking. In *Geography and Enlightenment*, 236–80. See Livingstone 1999.

———. 1999b. *Geography Unbound: French Geographic Science from Cassini to Humboldt.* Chicago: University of Chicago Press.

Gombrich, E. H. 1980. Standards of Truth: The Arrested Image and the Moving Eye. In *The Language of Images*, edited by W. J. T. Mitchell , 181–217. Chicago: University of Chicago Press.

Grady, John. 2006. Edward Tufte and the Promise of a Visual Social Science. In *Visual Cultures of Science*, 222–65. See Pauwels 2006.

Groenewegen, Peter. 2002. *Eighteenth-century Economics: Turgot, Beccaria and Smith and their Contemporaries.* London: Routledge.

Habermas, Jürgen. 1989. *The Structural Transformation of the Public Sphere: An Inquiry into a Category of Bourgeois Society.* Translated by Thomas Burger with the assistance of Frederick Lawrence. Cambridge, Mass.: MIT Press.

Harth, Erica. 1983. *Ideology and Culture in Seventeenth-Century France.* Ithaca: Cornell University Press.

Hayes, Julie Candler. 1999. *Reading the French Enlightenment: System and Subversion.* Cambridge: Cambridge University Press.

Hesse, Mary B. 1966. *Models and Analogies in Science.* Notre Dame, Ind.: University of Notre Dame Press.

Histoire, Sciences et Techniques: La revue. 2003. Catalogue of the exposition "Humboldt et Bonpland, 1799–1804, Une aventure savante aux Amériques, la boussole et l'orchidée." Paris: Musée des arts et métiers. 39/40.

Humboldt, Alexander von. 1811. *Essai politique sur le Royaume de la Nouvelle-Espagne.* 2 vols.

———. 1812. *Atlas géographique et physique du Royaume de la Nouvelle-Espagne fondé sur des observations astronomiques, des mesures trigonométriques et des nivellements barométriques.*

———. [1845] 1997. *Cosmos.* Translated by E. C. Otté. Introduction by Nicolaas A. Rupke. Baltimore: Johns Hopkins University Press.

Humboldt, Alexander von, and Aimé Bonpland. 1807. *Essai sur la géographie des plantes.* Rpt. 1977, New York: Arno Press.

Jardine, Nicholas. 1996. Naturphilosophie and the Kingdoms of Nature. In *Cultures of Natural History*, 230–45. See Jardine, Secord, and Spary 1996.

Jardine, N., J. A. Secord and E. C. Spary. 1996. *Cultures of Natural History.* Cambridge: Cambridge University Press.

La Condamine, Charles Marie de. [1745] 2004. *Voyage sur l'Amazone.* Edited by Hélène Minguet. Paris: La Découverte/Poche.

———. 1751a. *Journal du voyage fait par ordre du roi, à l'Équateur, servant d'introduction historique à la Mesure des trois premiers degrés du Méridien.*

———. 1751b. *Mesure des trois premiers degrés du Méridien dans l'Hémisphère Austral,*

Tirée des Observations de Mrs de l'Académie Royale des Sciences, Envoyés par le Roi sous l'Équateur.

Latour, Bruno. 1987. *Science in Action: How to Follow Scientific and Engineers through Society.* Cambridge, Mass.: Harvard University Press.

Livingstone, David N., and Charles W. J. Withers, eds. 1999. *Geography and Enlightenment.* Chicago: University of Chicago Press.

May, J. A. 1970. *Kant's Concept of Geography and its Relation to Recent Geographical Thought.* Appendix: A Translation of the Introduction to Kant's "Physische Geographie." Toronto: University of Toronto Press.

Miller, David Philip, and Peter Hanns Reill, eds. 1996. *Visions of Empire: Voyages, Botany, and Representation of Nature.* Cambridge: Cambridge University Press.

Minguet, Charles. 1997. *Alexandre de Humboldt: Historien et géographe de l'Amérique espagnole (1799–1804).* Paris: L'Harmattan.

Nicolson, Malcom. 1990. Alexander von Humboldt and the Geography of Vegetation. In *Romanticism and the Sciences,* edited by Andrew Cunningham and Nicholas Jardine, 169–85. Cambridge: Cambridge University Press.

———. 1995. Historical Introduction. Introduction to *Personal Narrative of a Journey to the Equinoctial Regions of the New Continent* by Alexander von Humboldt. London: Penguin Classics.

Pagden, Anthony. 1993. *European Encounters with the New World: From Renaissance to Romanticism.* New Haven: Yale University Press.

———, ed. 2000. *Facing Each Other: The World's Perception of Europe and Europe's Perception of the World.* The European Impact on World History 1450–1800, vol. 31. 2 vols. Aldershot, UK: Ashgate. Introduction, xvii–xxxvi.

Palsky, Gilles. 1996. *Des chiffres et des cartes: Naissance et développement de la cartographie quantitative française au XIXᵉ siècle.* Paris: Ministère de l'Enseignement supérieur et de la Recherche, Comité des travaux historiques et scientifiques.

Pauwels, Luc, ed. 2006. *Visual Cultures of Science: Rethinking Representational Practices in Knowledge Building and Science Communication.* Hanover, N.H.: Dartmouth College Press.

Picon, Antoine. 2004. The Freestanding Column in Eighteenth-Century Religious Architecture. In *Things That Talk: Object Lessons from Art and Science,* 101–46. See Daston 2004.

Porter, Roy. 1980. The Terraqueous Globe. In *The Ferment of Knowledge: Studies in the Historiography of Eighteenth-Century Science,* 285–324. Edited by G. S. Rousseau and Roy Porter. Cambridge: Cambridge University Press.

Playfair, William. [1787, 1801] 2005. *The Commercial and Political Atlas and Statistical Breviary.* Edited by and with an introduction by Howard Wainer and Ian Spence. Cambridge: Cambridge University Press.

Pratt, Mary Louise. 1992. *Imperial Eyes: Travel Writing and Transculturation.* London: Routledge.

Proust, Jacques. 1965. L'Encyclopédie. Paris: Armand Colin.

———. 1985. Marges d'une utopie: Pour une lecture critique des planches de l'Encyclopédie. Office du Livre en Poitou-Charente: Le temps qu'il fait.

Quesnay, François. [1758] 1894. Tableau Œconomique by François Quesnay. Facsimile edition. London: Macmillan.

Reill, Peter Hanns. 2005. Vitalizing Nature in the Enlightenment. Berkeley and Los Angeles: University of California Press.

Richards, Robert J. 2002. The Romantic Conception of Life: Science and Philosophy in the Age of Goethe. Chicago: University of Chicago Press.

Rorty, Richard, ed. and intro. 1967. The Linguistic Turn: Recent Essays in Philosophical Method. Chicago: University of Chicago Press.

Rudwick, Martin J. S. 1976. The Emergence of a Visual Language for Geological Science 1760–1840. History of Science 14:149–95.

———. 1978. Transposed Concepts from the Human Sciences in the Early Work of Charles Lyell. In Images of the Earth: Essays in the History of the Environmental Sciences, edited by L. J. Jordanova and Roy S. Porter, 67–83. The British Society for the History of Science.

———. 1996. Minerals, Strata and Fossils. In Cultures of Natural History, 266–86. See Jardine, Secord, and Spary 1996.

Rupke, Nicolaas A. 1996. Humboldtian Medicine. Medical History 40:293–310.

———. 2001. Humboldtian Distribution Maps: The Spatial Ordering of Scientific Knowledge. In The Structure of Knowledge, 93–116. See Frängsmyr 2001.

———. 2005. Alexander von Humboldt: A Metabiography. Frankfurt: Peter Lang.

Sachs, Aaron. 2003. The Ultimate "Other": Post-colonialism and Alexander von Humboldt's Ecological Relationship with Nature. History and Theory 42:111–35.

Schiller, Friedrich. [1795] 1954. On the Aesthetic Education of Man in a Series of Letters. Translated by Reginald Snell. New York: Frederick Ungar Publishing Co.

Stafford, Barbara Maria. 1984. Voyage into Substance: Art, Science, Nature, and the Illustrated Travel Account, 1760–1840. Cambridge, Mass.: MIT Press.

———. 1994. Artful Science: Enlightenment and the Eclipse of Visual Education. Cambridge, Mass.: MIT Press.

———. 1999. Visual Analogy: Consciousness as the Art of Connecting. Cambridge, Mass.: MIT Press.

Suaudeau, R. 1958. Les représentations figurées des physiocrates. Paris: Librairie du Recueil Sirey.

Todorov, Tzvetan. 1968. La poétique structurale. In Qu'est-ce que le structuralisme?, 99–166. Edited by Oswald Ducrot, Tzvetan Todorov, Dan Sperber, Moustafa Safouan, and François Wahl. Paris: Seuil.

Tresch, John. Forthcoming. Even the Tools Will be Free: Humboldt's Romantic Technologies. In *The Heavens on Earth: Nineteenth-Century Observatory Sciences*, edited by Charlotte Bigg and Davis Aubin. Durham, N.C.: Duke University Press.

Trumbo, Jean. 1996. Making Science Visible: Visual Literacy in Science Communication. In *Visual Cultures of Science*, 266–83. See Pauwels 1996.

Tufte, Edward R. 1982. *The Visual Display of Quantitative Information.* Cheshire, Conn.: The Graphics Press.

———. 1990. *Envisioning Information.* Cheshire, Conn.: The Graphics Press.

———. 1997. *Visual Explanations: Images and Quantities, Evidence and Narrative.* Cheshire, Conn.: The Graphics Press.

———. 2006. *Beautiful Evidence.* Cheshire, Conn.: The Graphics Press.

Vaihinger, Hans. 1935. *The Philosophy of 'As if': A System of the Theoretical, Practical and Religious Fictions of Mankind.* Translated by C. K. Ogden. London: Kegan Paul, Trench Trubner and Co.

Wainer, Howard. 1997. *Visual Revelations: Graphical Tales of Fate and Deception from Napoleon Bonaparte to Ross Perot.* New York: Springer Verlag, Copernicus.

———. 2005. *Graphic Discovery: A Trout in the Milk and Other Visual Adventures.* Princeton: Princeton University Press.

Wainer, Howard, and Ian Spence. 2005. Introduction. William Playfair, *The Commercial and Political Atlas and Statistical Breviary.* Cambridge: Cambridge University Press.

Werner, Stephen. 1993. *Blueprint: A Study of Diderot and the Encyclopedia Plates.* Birmingham, Ala.: Summa Press.

Wise, M. Norton, and Elaine M. Wise. 2004. Staging an Empire. In *Things That Talk: Object Lessons from Art and Science*, 101–46. See Daston 2004.

Wolter, John A. 1972. The Heights of Mountains and the Lengths of Rivers. *Journal of the Library of Congress* 29:187–206.

Plant Species Cited in
Humboldt's Essay and *Tableau physique*

Stephen T. Jackson

Some 354 different kinds of plants are mentioned in the text and tables of Humboldt's *Essay*. Most of these names are still in use, but many are obscure or archaic. Most are from the Andes or Europe, but Humboldt also mentions in passing many species from North America, Mexico, Africa, Asia, and Oceania. In an effort to help the reader navigate the *Essay*, I have assembled a table indexing all of Humboldt's names with their modern counterparts (synonymies), where known. The table also includes the regions of the world where the plant is found, the page number(s) in the *Essay* where the plant is mentioned, and the family for all vascular plants. For non-vascular plants, I listed only the general group (Fungi, lichen, etc.). This table is not intended to be authoritative. I have attempted to provide modern families and species synonymies as comprehensively and accurately as possible, but I have no doubt that some have been missed or misplaced along the way. In most cases, the absence of a modern synonym indicates that Humboldt's name is still in use. Please note that this table includes only plants mentioned in the text of the *Essay*; many additional species are portrayed in the *Tableau physique*.

Humboldt's Name	Page Number	Common Name	Region	Synonyms	Family or Group
Abies taxifolia[1]	96	Silver fir	Europe	Abies alba	Pinaceae
Acaena	92	Acaena	Southern Hemisphere		Rosaceae
Achillea atrata	95	Yarrow	Europe		Asteraceae
Achillea nana	95	Yarrow	Europe		Asteraceae
Aesculus flava	143	yellow buckeye	North America		Hippocastanaceae
Agaricus procerus	65	Parasol agaric, pasture agaric	Europe	Macrolepiota procera	Fungi (Basidiomycete)
Agrostis	93	Bentgrass	Worldwide		Poaceae
Agrostis alpina	95	Alpine bentgrass	Europe		Poaceae
Aira canescens	65	Gray clubawn grass	Europe	Corynephorus canescens	Poaceae
Alchemilla	92	Ladies' mantle	worldwide		Rosaceae
Alchemilla pentaphyllea	95		Europe	Alchemilla pentaphylla	Rosaceae
Allionia	88		North and South America		Nyctaginaceae
Alnus	75	alder	Eurasia, North America, Andes		Betulaceae
Alpinia	88		Asia, South America, Pacific islands		Zingiberaceae
Alsine media	68	star chickweed	Eurasia (introduced worldwide)	Stellaria media	Caryophyllaceae
Alstonia	93		worldwide tropics and subtropics		Apocynaceae
Alstroemeria	91	Peruvian lily	South America	Alstremeria; Alstraemeria	Alstroemeriaceae

Species	Page	Common name	Distribution	Synonym	Family
Anacardium caracoli	80, 131	caracoli; espave; espavel	South and Central America	*Anacardium excelsum*	Anacardiaceae
Androsace carnea	95	pink rock-jasmine	Northern Hemisphere		Primulaceae
Androsace chamaejasma	95	rock-jasmine	circumpolar	*Androsace chamaejasme*	Primulaceae
Androsace villosa	95		Eurasia, North Africa	*Androsace muscoidea*	Primulaceae
Annona chylimoya	134	cherimoya	South America	*Annona cherimola*	Annonaceae
Anthemis montana	96	mountain chamomile; dog-fennel	Europe	*Anthemis cretia*	Asteraceae
Anthericum liliago	65	St. Bernard's lily	Europe		Liliaceae
Apium graveolens	68	celery	Europe		Apiaceae
Arabis caerulea	95	blue rockcress	Europe		Brassicaceae
Arbutus madronno[2]	66	madroño; madrone	Mexico and SW USA	*Arbutus xalapensis*	Ericaceae
Arbutus uva ursi	143	bearberry; kinnikinnick	circumpolar	*Arctostaphylos uva-ursi*	Ericaceae
Aretia alpina	95		Europe	*Androsace alpina*	Primulaceae
Aretia vitaliana	95		Europe	*Vitaliana primulifera*; *Androsace vitaliana*	Primulaceae
Aristolochia sypho	143	Dutchman's pipe	North America	*Aristolochia sipho*	Aristolochiaceae
Arnica scorpioides	95		Eurasia	*Arnica montana* (?)	Asteraceae
Artemisia glacialis	95	glacier wormwood; alpine mugwort	Europe		Asteraceae
Artemisia rupestris	95	rock wormwood	Eurasia		Asteraceae
Arum	73, 92	arum	Eurasia, Africa		Araceae
Avena[3]	93	oats	Europe, Asia, Africa		Poaceae
Avicennia	87, 88	mangrove	worldwide tropics		Avicenniaceae
Azalea procumbens	95	dwarf azalea; alpine azalea	circumpolar	*Loiseleuria procumbens*	Ericaceae

(continued)

Humboldt's Name	Page Number	Common Name	Region	Synonyms	Family or Group
Bambusa[4]	67, 131	bamboo	tropical and subtropical Asia and Australia		Poaceae
Barnadesia	92		South America		Asteraceae
Befaria	80, 87	tarflower; fly catcher	North and South America		Ericaceae
Berberis	75, 92	barberry	worldwide		Berberidaceae
Boconia frutescens	91	tree poppy; parrot weed	South and Central America; Caribbean		Papaveraceae
Boletus botrytes	80		?	synonymy obscure	Fungi (Basidiomycete)
Boletus ceratophora	88	wood-rotting shelf fungus	worldwide (?)	Gloeophyllum odoratum	Fungi (Basidiomycete)
Bombax	87		tropical Asia, Africa, Australia		Bombacaceae
Bougainvillea	67		South America		Nyctaginaceae
Brathys	87	St. Johns wort	worldwide	Hypericum section Brathys	Clusiaceae
Brathys juniperina	65, 67		South America	Hypericum juniperina	Clusiaceae
Bromelia karatas	65		South and Central America	Bromelia humilis; B. agavifolia; B. plumieri	Bromeliaceae
Byssus speciosa	88		Worldwide	Hypha speciosa; Trentepohlia (?)	Chlorophyta (green algae)
Cactus[5]	73	cactus	North and South America	Opuntia	Cactaceae
Cactus pereskia	88	leaf cactus, rose cactus	South and Central America	Pereskia	Cactaceae
Caesalpinia	88		worldwide tropics and subtropics		Fabaceae
Cardamine alpina	95	alpine bittercress	Europe		Brassicaceae

Carex curvula	95	alpine sedge		Europe	Cyperaceae
Carex nigra	95	black sedge		circumpolar	Cyperaceae
Carica papaya	72	papaya		South and Central America	Caricaceae
Carolinea	65	pachira	*Pachira*	Neotropics	Bombacaceae
Caryocar	131			South America	Caryocaraceae
Cascarilla	90	cascarilla	*Croton eleuteria*	South and Central America	Euphorbiaceae
Castelleja integrifolia	93			South and Central America; Mexico	Orobanchaceae
Castilleja fissifolia	93			South America	Orobanchaceae
Castilloa	90	ule; Panama rubber tree; Mexican rubber tree	*Castilla*	South and Central America; Mexico	Moraceae
Cavanillesia	73			South and Central America	Bombacaceae
Cecropia peltata	88, 131	trumpet-tree; yagrumo hembra		South and Central America	Cecropiaceae
Cerastium lanatum	95	alpine chickweed	*Cerastium alpinum* ssp. *lanatum*	circumpolar	Caryophyllaceae
Ceroxylon andicola	87, 88	Andean wax palm	*Ceroxylon alpinum*	South America	Arecaceae
Chamaerops	88, 94	European fan palm		Europe and N. Africa	Arecaceae
Cheirostemon[6]	92	Mexican hand tree; devil's hand tree	*Chiranthodendron*	Mexico and Central America	Malvaceae
Chenopodium quinoa	72, 135	quinoa		South America	Chenopodiaceae
Chuquiraga	109			South America	Asteraceae
Cinchona	65, 89, 90, 91	cinchona		South and Central America	Rubiaceae

(continued)

Humboldt's Name	Page Number	Common Name	Region	Synonyms	Family or Group
Cinchona condaminea	89		South America	Cinchona officinalis	Rubiaceae
Cinchona cordifolia	89		South and Central America		Rubiaceae
Cinchona glandulifera	89		South America	Cascarilla glandulifera	Rubiaceae
Cinchona lancifolia	89		South America	Cinchona officinalis	Rubiaceae
Cinchona longiflora	89		South America	Exostema longiflorum	Rubiaceae
Cinchona oblongifolia	89		South America	Ladembergia riveroana	Rubiaceae
Cinchona officinalis	89	Peruvian bark; quinine bark	South America		Rubiaceae
Citrosma	92		South and Central America		Monimiaceae
Cladonia paschalis	65		worldwide (?)	Stereocaulon paschale	Lichen (fruticose)
Clavaria pistillaris	65	Club coral, pestle-coral mushroom	Europe, North America	Clavariadelphus pistillaris	Fungi
Clusia	93		tropical and subtropical Americas		Clusiaceae
Coccoloba	87	sea grape	tropical and subtropical Americas		Polygonaceae
Columella	93		South America	Columellia	Columelliaceae
Conocarpus	88, 131	button mangrove; buttonwood	worldwide tropics and subtropics		Combretaceae
Convolvulus brasiliensis	65, 88		Worldwide tropics	Ipomoea brasiliensis	Convolvulaceae
Convolvulus littoralis	88		worldwide tropics	Ipomoea littoralis	Convolvulaceae
Cortex angostura	87, 90	angostura bark; cusparia	South America	Galipea officinalis; Angostura trifoliata	Rutaceae

Crataegus aria	65	whitebeam	*Sorbus aria; Sorbus intermedia*	Europe	Rosaceae
Crocus multifidus	95			Europe	Iridaceae
Croton argentum	67		*Croton argenteum*		Euphorbiaceae
Crotum argenteum	65		*Croton argenteum*		Euphorbiaceae
Cupressus disticha	66	baldcypress	*Taxodium distichum*	North America	Cupressaceae
Cuspa	90	amargoso; cuspa	*Aspidosperma cuspa*	South America	Apocynaceae
Cusparia febrifuga	87	angostura	*Angostura*	South America	Rutaceae
Cyperus fuscus	65	brown flatsedge		Eurasia	Cyperaceae
Cypura	91		Synonymy obscure	unknown	Iridaceae?
Dactylis[7]	93	orchardgrass		Eurasia	Poaceae
Daphne	96			Eurasia, Africa	Thymelaeaceae
Daphne cneorum	97	rose daphne		Europe	Thymelaeaceae
Daphne laureola	97	spurge laurel		Eurasia	Thymelaeaceae
Daphne mezereum	97	mezereon		Eurasia	Thymelaeaceae
Datura	93	datura, angel trumpet, jimson weed		North and South America	Solanaceae
Dichondra	92			worldwide	Convolvulaceae
Dicranium scoparium	68	wind-blown moss	*Dicranum scoparium*	Worldwide	Bryophyta (mosses)
Dicranum	92			Worldwide	Bryophyta (mosses)
Dionaea muscipula	143	Venus fly-trap		North America	Droseraceae
Dorstenia	92			Worldwide tropics	Moraceae
Draba hirta[8]	95			circumpolar	Brassicaceae
Draba stellata	95			Europe	Brassicaceae
Dracontium[9]	73		*Symplocarpus*	Eurasia and North America	Araceae

(continued)

Humboldt's Name	Page Number	Common Name	Region	Synonyms	Family or Group
Duranta Ellisii	92	sky flower	South America		Verbenaceae
Duranta Mutisii	92		South America		Verbenaceaa
Embothrium	93		South America		Proteaceae
Embothryum emarginatum	93		South America		Proteaceae
Epidendrum[10]	73, 92	star orchid	South and Central America		Orchidaceae
Equisetum	74	horsetail; scouring rush	worldwide		Equisetales (horsetails)
Erica	96	heath	Eurasia		Ericaceae
Erica arborea	97	tree heath	Europe, Africa		Ericaceae
Erica tetralix	66, 97	cross-leaved heath	Europe		Ericaceae
Erica vagans	97	Cornish heath	Europe		Ericaceae
Erica vulgaris	66, 97	heath; heather	Europe, W. Asia	Calluna vulgaris	Ericaceae
Eriophorum alpinum	95	alpine cottongrass	circumpolar		Cyperaceae
Eriophorum Scheuchzeri	95	white cottongrass	circumpolar		Cyperaceae
Eroteum	92		South and Central America, Pacific Islands	Trichospermum	Tiliaceae
Erythroxylum peruvianum	135	coca; cocaine	South America	Erythroxylum coca	Erythroxyllaceae
Escallonia	75, 87, 93	escallonia; redclaws	South America		Escalloniaceae
Escallonia myrtilloides	65, 67, 80		South America		Escalloniaceae
Escallonia tubar	93		South America		Escalloniaceae
Espeletia frailexon	93, 109, 131	frailejón	South America		Asteraceae
Fagus sylvatica	96	European beech	Europe		Fagaceae

Species	Page	Common name	Distribution	Synonym	Family
Ficus	65, 90	banyan, fig	worldwide tropics		Moraceae
Fragaria elatior	68	musk strawberry	Europe	*Fragaria moschata*	Rosaceae
Fragaria vesca	65, 68	woodland strawberry	Northern Hemisphere		Rosaceae
Fragaria virginiana	68	common strawberry	North America		Rosaceae
Fritillaria meleagris	96	snakes-head fritillary	Europe		Liliaceae
Fuchsia	91		South and Central America; Pacific Islands		Onagraceae
Fucus	66, 88	kelp	worldwide (marine)		Phaeophyta (brown algae)
Fucus natans	65	sargassum weed, gulfweed	Atlantic Ocean (marine)	*Sargassum natans*	Phaeophyta (brown algae)
Fucus saccharinus	65	sweet kelp	worldwide (marine)	*Saccharina latissima*	Phaeophyta (brown algae)
Galium pyrenaicum	95	Pyrenean bedstraw	Europe		Rubiaceae
Genista lusitanica	96		Europe	*Echinospartum lusitanica*	Fabaceae
Gentiana	96	gentian	worldwide		Gentianaceae
Gentiana acaulis	97	stemless gentian	Europe		Gentianaceae
Gentiana campestris	97		Europe	*Gentianella campestris*	Gentianaceae
Gentiana ciliata	97	fringed gentian	Europe	*Gentianella ciliata*	Gentianaceae
Gentiana grandiflora	95		Eurasia (alpine)		Gentianaceae
Gentiana lutea	97	great yellow gentian	Europe		Gentianaceae
Gentiana nivalis	95	snow gentian	Europe		Gentianaceae
Gentiana pneumonanthe	97	marsh gentian	Europe, Siberia		Gentianaceae
Gentiana punctata	97	spotted gentian	Europe		Gentianaceae
Gentiana purpurea	95	purple gentian	Europe		Gentianaceae
Gentiana quitensis	93		South America		Gentianaceae
Gentiana verna	97	spring gentian	Europe		Gentianaceae

(continued)

Humboldt's Name	Page Number	Common Name	Region	Synonyms	Family or Group
Gleditschia	99	locust tree	North America, Asia	Gleditsia	Fabaceae
Gleditsia	73	locust tree	North America, Asia		Fabaceae
Gleditsia monosperma	143	water-locust; swamp-locust	North America	Geditsia aquatica	Fabaceae
Gleditsia triacanthos	143	honey locust	North America		Fabaceae
Glycine frutescens	143	American wisteria	North America	Glycine frutescens; Wisteria frutescens	Fabaceae
Godoya	67		South America		Ochnaceae
Gomphrena	101		Worldwide		Amaranthaceae
Gordonia Francklini[11]	143	franklinia; Franklin-tree	North America (Georgia)	Franklinia alatahama	Theaceae
Gunnera	92		South America; Pacific Islands, New Zealand		Gunneraceae
Gustavia augusta	80	gustavia; janiparindiba; chopé	South America		Lecythidaceae
Gymnodermea sinuata[12]	88		unknown	No modern synonym (devalidated)	Fungi(Basidiomycete?)
Heliconia	67, 73, 74, 88, 89, 131			tropical Americas, Pacific Islands, SE Asia	Heliconiaceae
Hevea	65, 90	rubber tree	South America		Euphorbiaceae
Hibiscus	73		Worldwide		Malvaceae
Hydrocotile	92	pennywort	Worldwide	Hydrocotyle	Apiaceae

Scientific name	Page	Common name	Distribution	Synonymy/Accepted name	Family
Hymenaea	88		South and Central America; Africa		Fabaceae
Hypericum baccatum	92		South America (?)	Synonymy obscure	Clusiaceae
Hypericum cayenense	92		South America (?)	Synonymy obscure	Clusiaceae
Jarava	67, 93		South America		Poaceae
Jatropha[13]	70	manioc, cassava	South America	*Manihot*	Euphorbiaceae
Jatropha manihot	72, 134	manioc, cassava	South America	*Manihot esculenta*	Euphorbiaceae
Juglans nigra	143	black walnut	North America		Juglandaceae
Juncus trifidus	95	highland rush; three-leaved rush	circumpolar		Juncaceae
Killinga	101	spikesedge; greenhead sedge	Worldwide	*Kyllinga*	Cyperaceae
Koelreutera[14]	91	golden raintree	Asia	*Koelreuteria*	Sapindaceae
Kyllingia	67	spikesedge; greenhead sedge	Worldwide	*Kyllinga*	Cyperaceae
Laurus persea	88	avocado	Central America	*Persea americana*	Lauraceae
Lecythis	131	monkey pot; coco	South and Central America		Lecythidaceae
Lichen haematomma	65, 66	bloodstain lichen		*Haematomma coccineum*	Lichens (crustose)
Lichen icmadophila	66	candy-drop lichen; fairy barf		*Icmadophila ericetorum*	Lichens (crustose)
Lichen verticillatus	88			*Cladonia verticillata*	Lichens (fruticose)
Linnea borealis	96	twinflower	Northern Hemisphere		Caprifoliaceae
Lobelia	90, 93		worldwide		Lobeliaceae
Lobelia nana	93		South and Central America, Mexico		Lobeliaceae

(continued)

Humboldt's Name	Page Number	Common Name	Region	Synonyms	Family or Group
Lychnis dioica	65	catchfly; red campion	Europe	Silene dioica	Caryophyllaceae
Lysianthus	91		South and Central America	Lisianthius	Gentianaceae
Lysianthus longifolius	65		Caribbean Islands	Lisianthius longifolius	Gentianaceae
Macrocnemum	68, 91		South and Central America		Rubiaceae
Mammea	134	mamey; sapote; mammey apple	South and Central America; West Indies		Clusiaceae
Matisia cordata	87	sapote	South America		Bombacaceae
Mauritia	67	mauriti palm; moriche palm; aguaje	South America		Arecaceae
Melastoma[15]	87, 91, 93		tropical Asia, Africa, Australia		Melastomataceae
Merendera bulbocodium	95		Europe	Merendera pyrenaica	Colchicaceae
Mimosa	73		tropical and subtropical Americas		Fabaceae
Mimosa inga	88	inga	South and Central America	Inga	Fabaceae
Mitchella repens	143	partridgeberry	North America		Rubiaceae
Mnium serpillifolium	67		Europe, Asia, North America	Mnium punctatum; Rhizomnium punctatum	Bryophyta (mosses)
Molina	67, 93	molina	South America	Dysopsis	Euphorbiaceae
Musa	73, 88	banana; plantain	worldwide tropics and subtropics		Musaceae

Mussaenda[16]	88		tropical Asia, Africa, Indian Ocean		Rubiaceae
Mutisia	92		South America		Asteraceae
Myosotis perennis	95	forget-me-not	Eurasia	Myosotis scorpioides	Boraginaceae
Myrica	90		worldwide except Australia and vicinity		Myricaceae
Narcissus bicolor	96		Europe		Amaryllidaceae
Nectandra sanguinea	80		South America		Lauraceae
Nelumbium luteum	143	yellow water lotus	North America	Nelumbo lutea	Nelumbonaceae
Nerteria	92	pincushion; coral bead	South and Central America; Australasia	Nertera	Rubiaceae
Nierembergia	92	cup flower	South America		Solanaceae
Ochroma	68, 73		South and Central America		Bombacaceae
Oxalis	92	wood sorrel	worldwide		Oxalidaceae
Panicum	93	panic-grass	worldwide		Poaceae
Papporophorum	67	pappusgrass	worldwide	Pappophorum	Poaceae
Passerina geminiflora	95		Europe (?)	Gnidia geminiflora (?)	Thymelaeaceae
Passerina nivalis	95		Europe	Thymelaea nivalis; Thymelaea tinctoria ssp. nivalis	Thymelaeaceae
Paullinia	73		tropical and subtropical Americas		Sapindaceae
Pinus	73	pine	Northern Hemisphere		Pinaceae
Pinus mugho	96	mountain pine	Europe	Pinus mugo	Pinaceae
Pinus strobus	65, 96	eastern white pine	eastern North America		Pinaceae

(continued)

Humboldt's Name	Page Number	Common Name	Region	Synonyms	Family or Group
Pinus sylvestris	65, 96	Scotch pine	Eurasia		Pinaceae
Plumeria	88	frangipani	South and Central America		Apocynaceae
Polygonum aviculare	65	knotweed	Eurasia		Polygonaceae
Polygonum bistorta	65	bistort	Eurasia	Persicaria bistorta	Polygonaceae
Polymnia	93		eastern North America		Asteraceae
Polytrichum commune	65, 68		Worldwide		Bryophyta (mosses)
Polytrichum piliferum	65		Eurasia, North America		Bryophyta (mosses)
Porlieria	73		North and South America		Zygophyllaceae
Porlieria hygrometrica	92		South America		Zygophyllaceae
Portulaca oleracea	68	purslane	Worldwide		Portulacaceae
Potentilla lupinoides	95		Europe		Rosaceae
Pothos[17]	73	devils ivy	tropical Asia	Epipremnum	Araceae
Primula	96	primrose	Worldwide (except Australia)		Primulaceae
Primula auricula	98	auricula; bear's ear	Europe		Primulaceae
Primula elatior	97	oxlip	Europe		Primulaceae
Primula integrifolia	97		Europe		Primulaceae
Primula farinacea	95, 98		Europe (?)	Synonymy obscure	Primulaceae
Primula farinosa	98	birds-eye primrose	Eurasia		Primulaceae
Primula marginata	98		Europe		Primulaceae
Primula villosa	97		Europe		Primulaceae
Primula viscosa	95, 98		Europe	Primula latifolia	Primulaceae

Species	Ref.	Common name	Species	Distribution	Family
Psychotria	65, 87	chacruna		worldwide tropics and subtropics	Rubiaceae
Quercus granatensis	92	Andean oak	Quercus humboldtii	South America (Colombia)	Fagaceae
Quercus robur	96	pedunculate oak		Eurasia	Fagaceae
Ranunculus Gusmani	93		Ranunculus guzmanii	South America	Ranunculaceae
Ranunculus	96	buttercup		Worldwide	Ranunculaceae
Ranunculus alpestris	95, 97	alpine buttercup		Europe	Ranunculaceae
Ranunculus amplexicaulis	97			Europe	Ranunculaceae
Ranunculus aquatilis	97	white water-buttercup; white crowfoot		Circumpolar	Ranunculaceae
Ranunculus glacialis	95, 97, 98	glacier buttercup		Europe	Ranunculaceae
Ranunculus gouani	96, 97			Europe	Ranunculaceae
Ranunculus nivalis	97			circumpolar	Ranunculaceae
Ranunculus parnassifolius	95, 97			Europe	Ranunculaceae
Ranunculus seguierii	98			Europe	Ranunculaceae
Ranunculus thora	97			Europe	Ranunculaceae
Rhizophora mangle	65, 88	red mangrove		coastal tropics	Rhizophoraceae
Rhododendrum	98	rhododendron; azalea	Rhododendron	worldwide	Ericaceae
Rhododendrum chamaecistus	98		Rhododendron chamaecistus	Europe	Ericaceae
Rhododendrum ferrugineum	96, 98	alpenrose	Rhododendron ferrugineum	Europe	Ericaceae
Rhododendrum hirsutum	98		Rhododendron hirsutum	Europe	Ericaceae
Ribes frigidum	93			South America	Grossulariaceae
Rubus	75	raspberry		worldwide	Rosaceae
Rubus saxatilis	96	stone bramble		Europe	Rosaceae

(continued)

Humboldt's Name	Page Number	Common Name	Region	Synonyms	Family or Group
Salix herbacea	95	dwarf willow	circumpolar		Salicaceae
Salix reticulata	95	net-leaved willow	circumpolar		Salicaceae
Salix retusa	95		Europe		Salicaceae
Saxifraga	96	saxifrage	Worldwide		Saxifragaceae
Saxifraga aizoon	97, 98	Saxifrage, aizoon	boreal Northern Hemisphere		Saxifragaceae
Saxifraga androsacea	95, 97, 98		Eurasia		Saxifragaceae
Saxifraga aspera	95, 98		Europe		Saxifragaceae
Saxifraga autumnalis	98		circumpolar	Saxifraga aizoides	Saxifragaceae
Saxifraga biflora	95		Europe		Saxifragaceae
Saxifraga bryoides	95, 98	moss saxifrage	Europe		Saxifragaceae
Saxifraga burseriana	98		Europe		Saxifragaceae
Saxifraga caesia	98		Europe		Saxifragaceae
Saxifraga cespitosa	97		Eurasia, North America		Saxifragaceae
Saxifraga cotyledon	98	silver saxifrage	Europe		Saxifragaceae
Saxifraga exarata	97		Eurasia		Saxifragaceae
Saxifraga geum	97		Europe	Saxifraga taygetea	Saxifragaceae
Saxifraga granulata	97		Europe, Africa		Saxifragaceae
Saxifraga groenlandica	95, 97	tufted alpine saxifrage	circumpolar	Saxifraga cespitosa	Saxifragaceae
Saxifraga longifolia	97		Europe, Africa		Saxifragaceae
Saxifraga moschata	98	musky saxifrage	Europe		Saxifragaceae
Saxifraga mucosa	98		Europe?	Synonymy obscure	Saxifragaceae

Species		Synonym	Common name	Distribution	Family
Saxifraga oppositifolia	95, 97		purple saxifrage	circumpolar	Saxifragaceae
Saxifraga petraea	98	Saxifraga rosacea		Europe	Saxifragaceae
Saxifraga pyramidalis	97	Saxifraga cotyledon		Europe	Saxifragaceae
Saxifraga stellaris	95		starry saxifrage	Europe	Saxifragaceae
Saxifraga tridactylides	97	Saxifraga tridactylites		Europe, Asia, Africa	Saxifragaceae
Saxifraga umbrosa	97			Europe	Saxifragaceae
Scutellaria alpina	95		alpine skullcap	Eurasia	Lamiaceae
Sempervivum arachnoideum	95		hen-and-chicks; cobweb houseleek	Europe	Crassulaceae
Sempervivum montanum	95		mountain houseleek	Europe	Crassulaceae
Senecio persicifolius	95	Senecio pyrenaicus		Europe	Asteraceae
Seratis	73	Serapias?		Europe	Orchidaceae
Sesuvium	101			worldwide	Aizoaceae
Sesuvium portulacastrum	65, 88, 89		tropical and subtropical cencilla; sea-purslane	Americas	Aizoaceae
Sibbaldia procumbens	95			Circumpolar	Rosaceae
Sida pichinchensis	93	Nototriche pichinchensis		South America	Malvaceae
Silene acaulis	95		moss campion	circumpolar	Caryophyllaceae
Sisyrinchium	91		blue-eyed grass	North and South America	Iridaceae
Solanum	70		nightshade	Worldwide	Solanaceae
Solanum dulcamara	65		bittersweet nightshade; climbing nightshade	Eurasia	Solanaceae
Solanum nigrum	68		black nightshade; Eurasia wonderberry		Solanaceae

(continued)

Humboldt's Name	Page Number	Common Name	Region	Synonyms	Family or Group
Solanum tuberosum	135	potato	South America		Solanaceae
Soldanella alpina	95	alpine snowbell	Europe		Primulaceae
Sonchus oleraceus	68	sowthistle	Eurasia		Asteraceae
Spermacoce	92		North and South America; Australia		Rubiaceae
Sphaeria digitata	65	dead man's fingers	Worldwide	Xylaria digitata, Xylaria polymorpha	Fungi (Basidiomycete)
Sphagnum	67		Worldwide		Sphagnophyta (peat mosses)
Sphagnum palustre	65, 66		Worldwide		Sphagnophyta (peat mosses)
Staehelina	93, 101		Mediterranean region		Asteraceae
Statice armeria	95	sea-thrift	Circumpolar	Armeria maritima	Plumbaginaceae
Sterculia	73	tropical chestnut	Worldwide tropics and subtropics		Sterculiaceae
Stipa	93	needlegrass	worldwide		Poaceae
Strelitzia	73	bird-of-paradise	Africa		Strelitziaceae
Swertia	101		Worldwide (except Australia)		Gentianaceae
Swertia quadricornis	93, 109		South America		Gentianaceae
Symplocos	92		Worldwide (Americas, Asia, Australia)		Symplocaceae
Talinum	88	fameflower	North and South America		Portulacaceae

Taxus	73	yew	Eurasia, North and Central America		Taxaceae
Taxus communis	96	European yew	Eurasia	Taxus baccata	Taxaceae
Tetraphis	92		Eurasia, North America		Bryophyta (mosses)
Thalinum	101	South America			Portulacaceae
Thelephora hirsuta	65	shelf fungus, bracket fungus	Eurasia, North America	Stereum hirsutum	Fungi
Theophrasta	88		South America		Theophrastaceae
Theophrasta americana	65		South America		Theophrastaceae
Thibaudia	91		South and Central America		Ericaceae
Thuia	99	white-cedar		Thuja	Pinaceae
Tourrettia	67		South and Central America		Bignoniaceae
Triticum	134	wheat	Eurasia		Poaceae
Triticum spelta	72	European spelt	Europe, W. Asia		Poaceae
Tussilago alpina	95	coltsfoot	Eurasia		Asteraceae
Tussilago farfara	95	coltsfoot	Eurasia		Asteraceae
Ulva	66, 68	sea-lettuce	worldwide (marine)		Chlorophyta (green algae)
Umbilicaria pustulata	94	rock tripe		Lasallia pustulata	Lichen (foliose)
Vaccinium myrtillus	65	bilberry	Europe		Ericaceae
Vallea stipularis	92		South America		Elaeocarpaceae
Verrucaria geographica	94			Rhizocarpon geographicum	Lichen (crustose)
Verrucaria limitata	68			Verrucaria cyanea	Lichen (crustose)

(continued)

Humboldt's Name	Page Number	Common Name	Region	Synonyms	Family or Group
Verrucaria sanguinea	68			Bellemerea sanguinea? Chiodecton sanguinea?	Lichen (crustose)
Viola biflora	95	yellow wood violet; two-flower violet		Circumpolar	Violaceae
Vites	71	grape	Europe, Asia, North America	Vitis	Vitaceae
Vitis	73	grape	Europe, Asia, North America		Vitaceae
Vitis vinifera	71	European grape; wine grape	Europe		Vitaceae
Weimannia	92		South America	Weinmannia	Cunoniaceae
Weissia	92		worldwide		Bryophyta (mosses)
Weissia paludosa	65		Europe, Asia, North America	Seligeria pusilla	Bryophyta (mosses)
Wintera[18]	87, 89, 90, 93		South and Central America	Drimys	Winteraceae
Wintera granatensis	80, 93	paramo pepper	South America	Drimys granatensis; Drimys winteri	Winteraceae

1. Synonymous with Pseudotsuga, but Humboldt discusses this species in context of the Pyrenees, so it must be Abies alba.

2. Synonymy is obscure. Humboldt was probably referring to Arbutus xalapensis, which grows in the regions of Mexico he visited and is locally referred to as madroño.

3. Humboldt may have been referring to a different grass not currently in this genus (which is not native to South America).

4. Humboldt was probably referring to bamboos that have been reassigned to other, related genera. Bambusa as such is not native to South America.

5. Humboldt refers specifically to the prickly-pears, which are now in genus Opuntia.

6. The Andean plant to which Humboldt refers is not clear. The genus Chirosthemon is no longer in usage.

7. This probably refers to a different genus; Dactylis is not native to South America.

8. Now split into several species.

9. Linnaeus' original genus *Dracontium* included the skunk-cabbage of Eurasia and North America, now reclassified to *Symplocarpus*. Humboldt may have had this familiar plant in mind for the reader. *Dracontium*, also in the Araceae, occurs in the Neotropics.

10. Humboldt's *Epidendrum* may have been a broader conception. *Epidendrum* originally included all epiphytic orchids.

11. This species has not been seen in the wild since the early nineteenth century, though it is cultivated widely in botanical gardens.

12. Humboldt was referring to a sterile form of fungus. Its identity has not been established, and his genus *Gymnoderma* has been rejected as invalid. The name is now used for a lichen genus that has no correspondence to Humboldt's fungus.

13. *Jatropha* still exists as a genus, but all Humboldt's references are to the cultivated manioc.

14. Humboldt's meaning here is not clear; the context indicates he is referring to a bryophyte.

15. Humboldt refers to *Melastoma* in South America. All South American species of *Melastoma* have been reassigned to other genera, though all are still in the Melastomataceae.

16. Humboldt must have been referring to something subsequently reassigned to another genus in the Rubiaceae; this genus does not occur in South America.

17. Humboldt may have been referring to lianas in general; the context suggests he refers to a growth form, not a taxonomic unit.

18. All South American species of *Wintera* have been reassigned to *Drimys*.

Instruments Utilized in Developing the *Tableau physique*

Stephen T. Jackson

Humboldt's *Essay* and his other writings contain an array of quantitative measurements—of spatial coordinates (latitude, longitude, altitude) as well as physical and chemical phenomena. The need for spatial precision was not as obvious in Humboldt's time as it is in ours. Modern science relies heavily on precise measurement of innumerable phenomena. Such precision for physical, chemical, and biological properties, however, was not universally accepted in the late eighteenth and early nineteenth centuries. Humboldt's obsession with precision and measurement followed a late eighteenth-century tradition of Lavoisier and other French scientists, which was still considered controversial. Lavoisier and his followers postulated that precise measurement of subtle variations in physical and chemical systems would reveal fundamentally important properties and principles. Others, particularly in England and Germany, viewed such variations as noise and precision a waste of time and resources.[1] In retrospect, Lavoisier and Humboldt were correct in principle, and sometimes in practice (e.g., Humboldt's geomagnetic measurements). However, many of Humboldt's instruments (e.g., the eudiometers) were not up to the task of detecting subtle variations; Humboldt's exacting efforts during his voyages often yielded meaningless numbers.

Humboldt took a large and diverse collection of instruments on his voyage to the Americas, with which he obtained many of the measurements used to compile the *Tableau physique*. He devotes several pages of the first chapter of the *Personal Narrative* to a detailed enumeration of his physical and astronomical instruments and their sources.[2] These instruments were state-of-the-art for his time, constructed of wood, glass, brass, and mercury, all packed in wooden cases. Humboldt, Bonpland, and their helpers hauled this arsenal for thousands of miles across Venezuela, the

1 This controversy is discussed in detail in Michael Dettelbach's 1999 essay, "The Face of Nature: Precise Measurement, Mapping, and Sensibility in the Work of Alexander von Humboldt."

2. Pages 34–40 in volume 1 of the translation by Helen Maria Williams (7 vols., 1814–29). The section describing the expedition's instruments was omitted from the translations by Thomasina Ross (1851) and Jason Wilson (1995).

Orinoco drainage, the Andes, and much of Mexico. This essay briefly discusses the instruments Humboldt used to obtain measurements for the *Tableau*, based on the "Physical Tableau" text itself, descriptions in the *Personal Narrative*,[3] and the general state of the art for physical measurement at the beginning of the nineteenth century.

Latitude, Longitude, and Elevation. Measurement of latitude using a sextant was straightforward at the time of Humboldt's voyage. Latitude determination was based on measurement of the angle of elevation of some celestial object (typically the midday sun) above the horizon. Humboldt had two sextants. The first was a high-performance Ramsden instrument with a ten-inch radius and high-magnification telescopes.[4] The second was a portable "snuff-box" sextant with a two-inch radius and weaker telescopes. Humboldt described this latter instrument as being especially convenient for making measurements on horseback or in a small boat.

Longitudinal determinations were routinely made using a chronometer, a high-precision portable clock that kept the time for a reference location (Paris in Humboldt's case). Longitude was calculated from the difference between local time and the time at the reference point. However, Humboldt frequently made more precise measurements astronomically by determining the local timing of astronomical events, particularly occultation of the moons of Jupiter. The event would be perceived as simultaneous across the globe, and tables of predicted times for specific reference points ("ephemeredes") could be used to determine longitude. This technique was proposed by Galileo in 1610 and perfected in the following century by Cassini and Rømer.[5] Humboldt also timed lunar eclipses and measured lunar distances (between the moon and the sun or certain stars) to estimate longitude (again using prescribed tables and formulas).

3. This essay does not include the many other instruments Humboldt took on the voyage, which included a microscope, surveying chains, a balance, a rain-gauge, various magnetometers, dipping-needles, and other geomagnetic equipment, evaporating pans, plant presses, dissecting tools, a small chemical laboratory (with glassware, reagents, etc.), galvanic equipment, and "a great number of small tools necessary for travelers to repair such instruments as might be deranged from the frequent falls of the beasts of burden."

4. Jesse Ramsden (1735–1800) was a prominent London instrument-maker. Ramsden had a reputation for designing and manufacturing high-precision instruments of various kinds. Among other things, he developed the Ramsden theodolite, a 200-pound monster of a surveying instrument. His theodolites were used in the late eighteenth-century Ordnance Survey of Britain, which served the basis for British maps until the mid-twentieth century. Humboldt carried Ramsden barometers, thermometers, and other instruments.

5. The history of methods for determining longitude is discussed for a general readership in Dava Sobel's *Longitude*.

Humboldt carried two chronometers. His primary one was by "Lewis Berthoud, No. 27,"[6] which had formerly "belonged to the celebrated Borda."[7] Berthoud's chronometer No. 9 is apparently the instrument that Humboldt took on his voyage.[8] Humboldt also carried a "demi-chronometer by Seyffert, serving for ascertaining the longitude at short distances."[9] He probably used this smaller, durable instrument during overland journeys when the Berthoud chronometer was likely packed and stowed for protection.

Humboldt's astronomical collection included a high-quality, three-foot achromatic telescope "by Dollond."[10] He also carried a smaller telescope which he could attach to tree trunks in the field.

Site-specific elevations were determined mainly by air-pressure estimates. Humboldt had two Ramsden barometers. Humboldt also measured the boiling-point of

6. Pierre Louis Berthoud (1754–1813) was France's leading maker of clocks and chronometers in the late eighteenth century.

7. *Personal Narrative*, vol. 1, p. 34, Williams translation. Jean-Charles de Borda (1733–99) was a prominent French physicist and military engineer who pioneered a number of navigational and cartographic methods, including the repeating-circle. Borda commanded a French warship during the American Revolution and was taken prisoner off Barbados following a severe battle with a superior British force. Borda's repeating-circle was the primary contender against Ramsden's theodolite in a classic English-French competition over which nation could build the more precise surveying instrument. The repeating-circle was far more portable than Ramsden's theodolite and easily its match in precision and accuracy.

8. Berthoud kept excruciatingly detailed records of his individual instruments, including their designs, repairs, and provenances. These have been summarized in a beautiful monograph by Jean-Claude Sabrier, *Longitude at Sea in the Time of Louis Berthoud and Henri Motel* (1993). Berthoud's No. 27 was made in 1793 for the Chevalier de Fleurieu. No. 26 was made in 1796 for the Chevalier de Borda, and was the first precision timekeeper made according to the Revolutionary decimal system. There are no records of either chronometer passing into Humboldt's hands. No. 26 was later owned by Alexis Bouvard (1767–1843), Director of the Paris Observatory. Berthoud's No. 9 was his first pocket marine watch, made in 1786 for Chastenet de Puységur. No. 9 was extensively tested by a committee at the Royal Observatory, leading to Berthoud's being awarded a lucrative prize for its design and accuracy. No. 9 was repaired by Berthoud and sold to Alexander von Humboldt in October 1798. In a December 1798 letter from Marseilles, Humboldt reported to Berthoud on its performance and closed by asking that "if you see our respected M. Borda express to him my gratitude for the happy advice that he gave me to acquire this superb chronometer." (All information from Sabrier; Humboldt quotation from 101–102.) The source of the discrepancy between Humboldt's *Personal Narrative* and Berthoud's records is not clear.

9. Johann Heinrich Seyffert (1751–1818) was a self-taught Dresden watchmaker.

10. Peter Dolland (1730–1820) was a London telescope-maker (and brother-in-law of Jesse Ramsden). Thomas Jefferson also purchased telescopes from Dolland.

water at various locations using "an apparatus by Paul,"[11] but he apparently did not use this as a means of elevational estimation.

Elevations of distant points were estimated using standard triangulation techniques, which were also applied to determining latitudes and longitudes in some cases. He had a variety of instruments for these purposes, including a repeating-circle by Le Noir[12] (used to calculate the angular or zenithal distance between objects in any plane), a theodolite by Hurter[13] (for measuring horizontal and vertical angles), a plane level, a quadrant, a graphometer (a simple form of theodolite), and a "lunette d'épreuve," which he used to measure small angles for distant points and to determine the progress of eclipses. Of these, the repeating circle, an instrument invented in the late eighteenth century by Borda, was the most precise. A pair of small telescopes mounted separately on independently rotating, concentric brass rings allowed the user to obtain a series of angular measurements from the same point with minimal error. Errors were further reduced by averaging the individual measurements within a series. Surveying techniques did not change appreciably for more than 150 years after Humboldt's voyage, although field instruments were made increasingly portable and compact.

Temperature. Humboldt's interest in temperature is obvious; he recorded thousands of temperature measurements during his travels. He carried a variety of thermometers of various sizes and scales, including a "thermometrical lead" for measuring water temperature at depth from a ship.

Hygrometry (Humidity). Hygrometry, the measurement of the moisture content of ambient air, was in a relatively primitive state in Humboldt's time. All available measures were based on the tendency of certain materials (wood, hair, catgut, bone, ivory, plant

11. Probably Théodore-Marc Paul (1760–1832), a prominent Geneva instrument-maker. Humboldt lists other Paul instruments (hygrometers, thermometers, magnetometer, cyanometer). It is possible some of these were made by Paul's father, Jacques Paul (1733–96), who made many instruments for Saussure and Deluc.

12. Etienne Lenoir (1744–1822), a skilled Parisian instrument-maker, built the repeating-circles designed by Borda, including those used in the great French meridional survey of Méchain and Delambre. In 1799, Lenoir crafted the archetypal meter-stick, the precisely machined platinum bar that served as the metric standard. Humboldt had other Lenoir instruments with him, including a dipping needle and a variation compass.

13. Humboldt's theodolite was built by Johann Heinrich Hurter (1734–99), a Swiss-born artist and instrument-maker who spent much of his later career in London. Hurter was a popular and well-known portrait artist and miniaturist; many of his enamels and miniatures can be seen in museums in England and elsewhere in Europe. He was also a skilled metal-worker, preparing gilt frames for his enamels and miniatures. These skills undoubtedly prepared him for his parallel career as a maker of scientific instruments after 1780.

fibers) to swell and shrink as they respectively absorbed or lost moisture to the atmosphere. Most could not be related directly to estimates of moisture content or relative humidity. Humboldt used a formula developed by Jean d'Aubuisson de Voisins (1762 [1769?]–1841), another of Werner's students) to estimate vapor content of the air from his hygrometric measurements.

Humboldt carried two hygrometers, both crafted by Paul in Geneva. The Saussure hygrometer used a human hair of approximately 12 inches length, fixed at one end with the other end suspended over a small cylinder and counterweighted with approximately 200 mg. The cylinder rotates upon an axis with a scaled dial attached. Contraction and lengthening of the hair under dry and wet conditions, respectively, rotates the cylinder and, accordingly, the dial. The scale runs from 0 (absolutely dry) to 100 (saturated). The relationship between the Saussure scale and modern relative humidity is logarithmic, and so the device is most accurate at moderate to high humidities.[14]

The other is described as a whalebone hygrometer of Deluc. This could refer to one of two Deluc designs. Deluc's earlier (1773) hygrometer used a thin cylinder of whalebone or ivory, filled with mercury, and fitted with a scaled glass thermometer-like tube. Under moist conditions, water absorption by the ivory would expand the diameter of the cylinder and the mercury in the tube would drop; in dry air evaporative loss would contract the ivory, driving the mercury up the tube. Obviously, this hygrometer required corrections for ambient temperature. Deluc's second (1791) design was similar to the hair hygrometer, except it used a very thin strip of whalebone instead of hair. Unlike Deluc's first design, it recorded lengthening and shortening of the whalebone rather than lateral expansion and contraction. It seems likely that Humboldt took this second, more portable Deluc instrument on his voyage.

Deluc and Saussure quarreled over the relative merits of their designs. Humboldt discussed his experience with the two hygrometers in the *Personal Narrative* (vol. 6, pp. 786–88, Williams trans.) with characteristic diplomacy. The Saussure instrument performed well under a variety of circumstances and served as the standard for Humboldt's measurements. He noted, however, that the Saussure instrument was unsuited for use in windy conditions. The Deluc hygrometer was especially sensitive to fine variations in humid conditions, but response-time was so slow in dry climates that Humboldt was "often uncertain whether we have not ceased our observations before the instrument has ceased its movement."

Electrometry. In modern usage, electrometry consists of measuring electrical charge or potential, and electroscopy consists of determining the existence of charge. Humboldt had two "electrometers," but one was actually an electroscope. The Ben-

14. Saussure's hair hygrometer was still in widespread use in the first half of the twentieth century, and was used in early aviation. F. J. W. Whipple's 1921 essay discusses the physical basis for Saussure's hygrometer and its relationship with relative humidity.

net electroscope, designed by Rev. Abraham Bennet in the 1780s, consisted of two leaves of gold foil, suspended from a conductor (usually brass) inside a glass jar. In the presence of strong static-electricity fields, the foil leaves repel each other. This device is essentially the same as the commercial or homemade electroscopes used in basic science-teaching.

Saussure's electrometer was the first instrument that actually measured charge quantitatively. Saussure's instrument was similar to the Bennet device, with two exceptions. First, instead of metal foil, the device consisted of two balls of elder (*Sambucus*) pith suspended on strings inside a glass jar. Second, the glass was marked with a scale so the distance between the pith-balls could be measured. Saussure's instrument was originally an electroscope with an ordinal scale—distance between the balls is not linearly related to the intensity of the charge. The discovery of Coulomb's Law in 1784 led to direct estimation of electrical charge using the calibrated device.

Blueness of the Sky. The color of the sky was determined using a "cyanometer by Paul," a set of colored tiles ordered in a gradation from a deep, dark blue (1) to a very pallid, near-white hue (16). Cyanometry was based on a scale developed in the Alps by Saussure, and Humboldt's cyanometer was calibrated against de Saussure's standards.

Chemical Composition of the Atmosphere. Humboldt's primary concern was whether the "purity of air"—the oxygen content of the atmosphere—decreased with elevation above sea level. Several eudiometers had been designed in the late eighteenth century, starting with one by Joseph Priestley. All were based on volumetric displacement following oxygen absorption by chemical reaction or combustion. Humboldt took two eudiometers on his voyage. One, "of Fontana,"[15] used nitric oxide; the other "by Reboul" used phosphorus. Both were imprecise in practice, and Humboldt's measurements proved useless. Humboldt's *Tableau physique* and discussion were based on measurements in Europe (mountain and balloon ascents), using better instruments designed by Volta (e.g., his "electric pistol") and Gay-Lussac.

Gravity. Point-measurements of the strength of Earth's gravitational field were made in Humboldt's time using a standard pendulum. In the *Personal Narrative*, Humboldt lists an "invariable pendulum" made in Madrid among his instrument collection. However, he appears not to have used it in during his extended field excursions. He specifically mentions using his "inclination meter" in the "Physical Tableau," yielding an uncalibrated series of gravity estimates at various points in the Andes. It is not clear which of the instruments in the *Personal Narrative* he refers to; evidently one of his surveying instruments was doubling as a field-portable pendulum.

15. Abbé Felice Fontana (1730–1805), Florentine chemist and physicist.

Biographical Sketches

Stephen T. Jackson

More than ninety individuals are named in Humboldt's *Essay*. Most of these names will be obscure to all but the specialist or historian today, but they include some of the leading scientists and naturalists of the eighteenth and early nineteenth centuries. Many of these individuals were personal acquaintances or correspondents of Humboldt. Following are brief biographical notes for most of these individuals. I have omitted several, including his draftsmen (Schoenberger, Turpin), his instrument-makers (Bennet, Leslie, Paul), and peripheral figures (mainly cited as sources for barometric or other data). The draftsmen and some of the instrument-makers are briefly described in my accompanying essays on the *Tableau physique* and Humboldt's instrument collection. I have added six men who played important roles in Humboldt's botanical thinking (J. R. Forster, Goethe) or his voyage (Jefferson, Montúfar), or were important in Humboldt's subsequent scientific pursuits related to the *Essay* (Arago, Kunth). I have also added a sketch for Helen Maria Williams, who, as Humboldt's English translator in the years following his voyage, was his primary conduit to English-speaking readers.

François **Arago** (1786–1853)

François Arago was Humboldt's most intimate friend during his years in Paris. He was trained in Paris as a mathematician by Siméon Poisson, and in 1805 was selected to continue the meridional survey initiated by Delambre and Méchain. He was joined in this by Jean Baptiste Biot. The survey took him through Spain in the middle of the Peninsular War, where a Frenchman with instruments would naturally be under suspicion of spying. He was imprisoned, escaped, recaptured, and then released, only to be captured again in the Mediterranean and imprisoned in Algiers. He was eventually released and made his way back to Paris in 1809, his survey notebook intact. Arago was given an appointment at the Paris Observatory, which he held for the rest of his life. Arago conducted research on a variety of physical topics, including magnetism, sound, and meteorology, but his most important contributions were in the physics of light. He had strong liberal/republican political views, and in his political career in the 1830s and 1840s (Chamber of Deputies, Minister of Marine, Minister of War, President of the Executive Power Commission in the short-lived Second Republic) he advanced many reforms, including universal suffrage (for men), abolition of slavery in the French colonies, and abolition of flogging in the Navy. He also fostered development of the French railway and telegraph system, improved government funding for the sciences, and acquired the museum at Cluny. Arago's political career ended with his resignation from the Commission during General Cavaignac's brutal repression of riots in Paris. He was briefly arrested as an "enemy of the regime" following Napoleon III's coup d'état of 1851, and he afterwards refused to swear allegiance to the new emperor. His age, stature, and poor health prevented retribution. Arago kept a regular and lively correspondence with Humboldt until his death.

Benjamin Smith **Barton** (1766–1815)   (pp. 99, 143)

Barton, a Philadelphia physician, was a pioneering naturalist and scholar of Native American cultures. He became Professor of Botany at the University of Pennsylvania and was a member of the American Philosophical Society. A friend of Thomas Jefferson, Barton was asked by Jefferson to provide training in botany, zoology, and Indian culture to Meriwether Lewis in preparation for the expedition to explore the newly acquired Louisiana Purchase territory. Barton's 1803 book, Elements of Botany, was part of the small library that accompanied Lewis and Clark on their journey to the Pacific coast. Humboldt and Bonpland visited Barton in Philadelphia during their 1804 visit to the United States.

Claude-Louis **Berthollet** (1748–1822)   (p. 114)

Berthollet was a prominent French chemist, who played a major role in developing the system of chemical nomenclature proposed by Lavoisier. That system is still in use today. He also did pioneering work in physical chemistry, researched dyes and bleaches, and identified the chemical composition of ammonia. He was a friend of Napoleon, who invited him to accompany the campaign in Egypt, where he studied sodium carbonate (trona) deposition in arid lakes. Humboldt named the brazil-nut, Bertholletia excelsa, in his honor.

Jean Baptiste **Biot** (1774–1862)   (pp. 62, 111, 115, 119)

Biot was a prominent mathematician, astronomer, and physicist, based in Paris. His published works covered the entire range of applied mathematics, physics, geodesy, and astronomy. His most lasting contributions were his studies of the optical properties of gases, liquids, and crystalline solids. He worked out the principles of optical polarization and double refraction of light. Biot accompanied Gay-Lussac in his celebrated 1804 balloon ascent to 4,000 meters.

Aimé **Bonpland** (1773–1858)   (pp. 62, 68, 86, 87, 106, 128, 146)

Aimé Bonpland was Humboldt's traveling companion during the 1799–1804 voyage to the Americas. Trained as a naval surgeon in Paris, Bonpland's interest in botany was stimulated by his contact with Lamarck, Desfontaines, and Jussieu at the Jardin des Plantes. He met Humboldt in Paris in 1795, and they soon decided to attempt a voyage of scientific exploration together. Although Humboldt was an excellent botanist in his own right, Bonpland's specific interests in botany and skills as a field botanist led to his taking the lead botanical role in their explorations. Bonpland collected more than 60,000 plant specimens, often risking his life and health to obtain them. After their return to France, Bonpland was appointed director of the Empress Josephine's private botanical garden at Malmaison. Bonpland was a much better field botanist and horticulturist that he was a methodical desk scientist; Humboldt's hopes for Bonpland's preparation of botanical monographs and detailed descriptions of the plant collections were never fulfilled. Despite this, they remained close friends, and Humboldt recruited Carl Sigismund Kunth to finish the work. Following Josephine's death in 1814, Bonpland returned to South America, emigrating to Buenos Aires in the new Argentine Republic. After a few years there teaching and practicing medicine, Bonpland developed a plantation along the Paraná River on the Argentine/Paraguayan border, where he performed ambitious agricultural experiments and developed new varieties of several cultivars. In 1821 Bonpland was kidnapped by Paraguayan soldiers, who killed all of his servants and assistants. He was packed off to Paraguay and kept prisoner there for the next seven years by the paranoid dictator, Dr. Francia. Humboldt and other friends in Europe mounted a campaign for his release, which ultimately succeeded in 1829 after Francia grew tired of the international pestering. Bonpland developed new plantations in the Corrientes region of northwest Argentina. Excepting interruptions by a civil war from 1838 to 1842, he spent the rest of his life there in happiness, practicing medicine, collecting and cultivating plants, and raising a family with a native woman.

Pierre **Bouguer** (1698–1758)   (pp. 83, 83n, 103, 105, 116, 118, 129, 136, 137n, 142, 146)

Bouguer, a Breton mathematician and physicist, was, with Godin and La Condamine, one of the leaders of the quarrelsome 1735–45 expedition to South America in order to measure a longitudinal degree near the equator. In Peru and Ecuador, he made extensive barometric measurements as well as estimates of atmospheric refraction. He is best known for pioneering formal, systematic approaches to photometry, measurements of the relative brightness of celestial bodies, and estimates of light absorption by the atmosphere and by ocean waters.

Christian Leopold Freiherr von **Buch** (1774–1853) (pp. 126n, 139, 140)

Leopold von Buch was a contemporary of Humboldt's at the School of Mines in Freiberg. Buch became an influential geologist, whose studies of volcanic rocks led him to take a leading role in overturning the Neptunist notions of Werner, his mentor. Buch was eccentric, irritable, and unsociable, but he developed a lasting friendship with Humboldt, and they remained in close contact until Buch's death.

Georges-Louis Leclerc de **Buffon** (1707–88) (p. 85)

Buffon was the leading French natural historian of the eighteenth century. His *Histoire naturelle, générale et particulière,* consisting of thirty-six volumes, with another eight posthumous volumes, attempted to summarize everything known about the animal, plant, and mineral world. It was translated into many languages; the English translations were especially popular in England and the United States and were reprinted into the mid-nineteenth century. In 1739 Buffon was appointed *intendant* of the Jardin du Roi (now the Jardin des Plantes), which he developed from a curiosity garden into a major research institution. He spent the rest of his life there. Buffon's scientific accomplishments were diverse. He laid foundations for modern ecology, evolutionary biology, anthropology, and forestry, among others, and is credited with discovering the binomial theorem and introducing calculus into probability theory. Buffon challenged the biblical view of earth history and calculated that the earth was at least 75,000 years old, based on the cooling rate of iron. This led to his censure by the Roman Catholic Church and widespread burning of his books.

Francisco José **Caldas** (1768–1816) (p. 120)

Francisco Caldas, a self-taught naturalist and mathematician, was a native of Popayán in Colombia. Trained in the law, he read avidly the few scientific works he could obtain. Humboldt was greatly impressed by Caldas's knowledge and talents when they met. Caldas was taking accurate astronomical measurements from instruments he built himself from local materials, and was measuring altitude using the ingenious approach of recording the boiling-temperature of water. Humboldt confirmed many of the findings and methods that Caldas had developed entirely on his own, and brought him up to date on European physics and astronomy. They spent several weeks together in the vicinity of Quito and Chillo, collecting plants, measuring altitudes, dissecting a llama, climbing to the 4,800-m summit of Pichincha, and discussing a range of scientific topics. Caldas taught Humboldt a great deal about the geography and climate of the region, and provided him with his measurements of altitudes of various points. In turn, Humboldt passed on his ideas concerning patterns and controls of plant distributions, and both Humboldt and Bonpland provided Caldas with formal training in botany. After Humboldt's departure from Quito, Mutis appointed Caldas to the Botanical Expedition of Nueva Granada. Over the next few years, Caldas developed a series of biogeographical maps and monographs of the region. These documents were never published, and most appear to have been lost. He also amassed a large herbarium and measured elevations of many Andean landmarks. Caldas moved to Bogotá in 1805 to take the directorship of the new Astronomical Observatory, which Mutis was responsible for developing. Caldas, with many other scientists from the Botanical Expedition and Astronomical Observatory, took part in the independence movement in Colombia. Caldas, along with several of his colleagues, was executed by firing squad in Bogotá on General Morillo's orders. Caldas asked that his execution be postponed to allow him to finish some scientific works, specifically his phytogeographical studies, but Morillo refused, saying "Spain has no need of savants."

Augustin Pyramus de **Candolle** (1778–1841) (pp. 62, 96)

Candolle, a Swiss botanist, alternated between Paris, Montpellier, and Geneva as his base. His aptitude was recognized early on by Cuvier and Lamarck, who fostered his career. He conducted a botanical survey of France for the government and published widely on botanical subjects. His greatest contributions were in plant taxonomy (a term he coined). He also did important work in plant geography. His systematic and phytogeographical work was continued by his son, Alphonse, whose 1855 *Géographie botanique raisonnée* was strongly influenced by Humboldt's *Essay*.

Antonio José **Cavanilles** (1745–1804) (p. 90)

Cavanilles was one of Spain's leading botanists in the eighteenth century. He served as a priest in Paris and Madrid, and spent the last years of his life as director of the Royal Botanic Gardens in Madrid and as a professor of botany. He described many new plants from Australia and other parts of the South Pacific brought back from Malaspina's expedition of 1793. Humboldt and Bonpland visited Cavanilles in Madrid during their stay in 1799, just before departing for South America.

Vicente **Cervantes** (1755–1829) (p. 92)

Cervantes, a Spanish botanist and pharmacist, traveled to Mexico to take part in the Mexican botanical surveys of Sessé and Mociño. He remained in Mexico City to teach and research in botany and pharmacognosy. In recognition of his services to Mexico, Cervantes was exempted from Iturbide's 1821 decree expelling Spaniards from the newly independent "Empire of Mexico." Humboldt and Bonpland examined many of Cervantes's plant specimens and drawings while in Madrid in 1799.

Cosme Damián de **Churruca** y Elorza (1761–1805)　　　　(p. 130)

Churruca's life alternated between active service as an officer in the Spanish Navy and teaching mathematics, astronomy, and navigation. He did extensive mapping of the Straits of Magellan and other parts of Spanish South America, the Caribbean, and the Canary Islands. Some of these surveys were cut short when Churruca was called into action as Spain entered the various conflicts of the late eighteenth and early centuries. He died commanding a Spanish ship at the Battle of Trafalgar.

Gabriel **Ciscar** (1759–1829)　　　　(p. 142)

Ciscar taught mathematics and navigation from 1783 to 1798 at the Spanish Naval Academy in Cartagena. He introduced the metric system to Spain in 1800 while serving in the marine artillery as a senior officer. He was caught up in the turmoil associated with Napoleon's invasion of Spain and the resulting Peninsular War. Ciscar played an active role in the resistance and in drafting a liberal constitution for Spain. In 1814, following the restoration of Fernando VII to the throne, Ciscar was arrested and imprisoned until the revolution of 1820. He returned to government service, but was forced to flee to Gibraltar in 1823 when Fernando yet again revoked the constitution and repressed dissent. Ciscar spent the rest of his life in exile on Gibraltar, writing works on mathematics and astronomy.

Philibert **Commerson** (1727–73)　　　　(p. 98)

Commerson was a French naturalist who accompanied Louis de Bougainville's globe-girdling exploration (1766–69) of the Southern Hemisphere. He collected plants in South America and the archipelagoes of the South Pacific and Indian Oceans. When the expedition reached Ile de France (now Mauritius, Commerson disembarked, hoping to explore nearby Madagascar. He died shortly after his trip to Madagascar.

Antonio de **Córdoba** (1740?–1811)　　　　(p. 88)

Admiral Córdoba led two Spanish surveying expeditions to the Straits of Magellan between 1785 and 1788, and wrote the first comprehensive description of the lands and waters of Patagonia and Tierra del Fuego.

Manasseh **Cutler** (1742–1823)   (p. 99)

Cutler, a Congregationalist clergyman, served as a chaplain in the American Revolution, and towards the end of the war he acquired skills as a physician. After the war he developed an interest in botany and conducted the first systematic survey of the New England flora. He was part of the first party to ascend Mount Washington, the highest summit in New England. Cutler wrote the chapter on trees and plants for Jeremy Belknap's *History of New Hampshire* (1791–92), a work evidently consulted by Humboldt. He also published articles on astronomy, agronomy, and archeology. He helped settle Marietta, Ohio, in 1787, but shortly returned to Massachusetts, where he served in the state legislature and in the U.S. Congress.

Georges **Cuvier** (1769–1832)   (p. 126)

Cuvier, a zoologist based at the Jardin des Plantes and the Muséum national d'Histoire naturelle in Paris, was influential as a comparative anatomist in his time. However, his most lasting accomplishments were in historical geology and paleontology, a field that during his lifetime developed from a backwater concerned with curiosities into a formalized discipline focused on understanding the history of life on earth. Cuvier's 1796 discourse on living and fossil elephants demonstrated that now-extinct species had occurred on earth in the recent past. By the time of Humboldt's return to Paris, Cuvier had demonstrated the existence of numerous extinct mammals in the fossil record, and was developing his ideas that the earth had experienced a number of "revolutions" in the past, each leading to broad-scale extinctions and biotic transformations. Although he opposed the notion of organic evolution, noting the anatomical similarities between mummified ibises from ancient Egypt and their modern counterparts, his work in functional morphology and paleontology played a crucial role in the later development of evolutionary biology. Cuvier was politically adroit, staying in the good graces of revolutionary, imperial, and monarchical governments in his adult lifetime, and arranging to have the final word on colleagues and competitors by writing their official obituaries.

John **Dalton** (1766–1844)   (p. 109)

Dalton was from a Quaker family and thus educated outside the "mainstream" in England; Cambridge and Oxford were open only to Anglicans at the time. He spent his entire career in Manchester, teaching and tutoring in mathematics and chemistry. His initial scientific interests, persisting through his lifetime, were in meteorology. Humboldt's reference in the *Essay* is to Dalton's meteorological work. Dalton's research into the physical properties of atmospheric gases led to his development of the atomic theory and the concept of atomic weights, which remain as foundation principles of chemistry.

Jean-Baptiste-Joseph **Delambre** (1749–1822)   (pp. 62, 103, 103n, 118, 141, 141n, 142)

Delambre, based at the Paris Observatory and the Collège de France, was an astronomer with particular mathematical talents. In 1792, he published detailed calculations of the predicted orbits of Jupiter, Saturn, and Uranus. Uranus, however, was soon observed to be moving faster and then slower in its orbit than Delambre predicted. This anomaly ultimately led to the discovery of the planet Neptune, one of the great achievements of Newtonian science. Delambre was also heavily involved in geodesy, and was a leader in the 1790s effort to estimate the precise length of the earth's meridian, the basis for the metric system of measurement. His efforts were interrupted by his arrest during the revolutionary period because his field instruments led to suspicions that he was a spy. Eventually he was able to continue the work, and published a three-volume monograph, *Base du système métrique* (1806–10).

Jean-André **Deluc** (1727–1817)   (pp. 107, 109, 110, 139, 140)

Deluc spent much of his adult life, until his mid-forties, as a businessman and naturalist in his native Geneva. Following a major business failure in 1773, he emigrated to London, where he was given an appointment as reader to Queen Charlotte, providing both income and time to pursue his scholarly interests. He soon published an influential volume (dedicated, of course, to Charlotte) advocating a catastrophic account of geologic history consistent with the biblical account in Genesis. A generation later, Cuvier set aside Deluc's scriptural arguments but retained the catastrophic ones in developing his idea of global revolutions. Deluc was a highly original thinker in meteorology, conducting a variety of ingenious experiments on temperature, moisture, and evaporation. He designed portable barometers for use in the field and refined the formulas necessary to measure altitudes barometrically. He also invented a novel hygrometer based on the hygrometric expansion/contraction of ivory. Humboldt used Deluc's barometers and hygrometer during his explorations.

René-Louiche **Desfontaines** (1750–1833)   (p. 59)

René Desfontaines studied plant taxonomy, anatomy, and physiology at the Jardin du Roi in Paris (now the Jardin des Plantes). He conducted botanical explorations of the Atlas Mountains and other regions of northwest Africa, describing hundreds of new species. In 1786 he was given a lifetime appointment as Professor of Botany at the Jardin du Roi and the Muséum national d'Histoire naturelle. Desfontaines' research accomplishments and influence never matched those of his colleague and contemporary, Antoine Laurent de Jussieu, but he was a strong teacher. His public lectures on botanical subjects, delivered weekly at 7 a.m., were immensely popular, attracting hundreds.

Louis Benjamin **Fleuriau** de Bellevue (1761–1852)   (p. 105)

Fleuriau de Bellevue was born to a prosperous Huguenot family from La Rochelle; his father made a fortune in the lucrative Caribbean sugar trade. After studying in Geneva, he traveled extensively in southern Europe, often as a protégé of the geologist Déodat de Dolomieu (for whom dolomite was named). Fleuriau became a prominent mineralogist and geologist in the early nineteenth century. Fleuriau had several mulatto half-siblings from his father's extended sojourn in Port-au-Prince, Saint-Domingue (now Haiti), some of whom were schooled in La Rochelle. All his relatives who remained in or returned to Haiti were killed in the turmoil of the Haitian Revolution.

George **Forster** (1754–94)   (p. 61)

George Forster learned cartography and natural history from his father, Johann Reinhold Forster, whom he accompanied on a survey of parts of Russia at the age of ten. The Forsters relocated from Prussia to England in 1766. George accompanied his father on James Cook's second expedition to the Pacific (1772–75) and published a narrative account of the voyage, with natural history and anthropology interspersed throughout, in 1777. This book is still valued for its objective and sophisticated observations of Polynesian and other Pacific societies. Forster returned to Germany in 1778, where he taught in Cassel and Mainz. Humboldt and Forster were introduced by Forster's father-in-law, one of Humboldt's professors in Göttingen. They became close friends, and Humboldt accompanied Forster on a trip through the Low Countries, England, and France. Forster later published an account of this journey. Forster was a strong sympathizer with the republicans in the French Revolution, and when the revolutionary army seized Mainz in 1792, Forster became one of the leaders of the short-lived, democratic Mainz Republic. Forster was in Paris as an envoy when the Prussian and Austrian armies captured Mainz. Declared an outlaw by the Holy Roman Emperor, Forster died of an illness while in exile in Paris. George Forster's influence on Humboldt was extensive. Forster integrated science and aesthetics in his writings, viewing the world as something to be analyzed and embraced simultaneously. Humboldt pursued this vision through his lifetime, with a final synthesis in *Cosmos*. Forster's account of his transglobal voyage stimulated Humboldt's interest in travel, and his view that landscape topography and vegetation shaped development of human societies was adopted and expanded by Humboldt.

Johann Reinhold **Forster** (1729–98)

Johann Forster was a Lutheran pastor and naturalist whose family emigrated to Prussia from Scotland during Cromwell's rule. He was commissioned by Tsarina Catherine II to conduct a survey of colonies along the Volga. Following a dispute with the Russian bureaucracy (resulting from his outspoken criticisms of the administration of the provinces he visited), Forster moved to England where he succeeded Joseph Priestley at Warrington Academy. Having developed a reputation as a naturalist, Forster was invited to serve as naturalist on James Cook's second voyage to explore the Southern Hemisphere. Between 1772 and 1775, Forster traversed the South Pacific with Cook, making observations of the natural history and ethnology of the various islands they visited. Forster's account of the voyage, published in 1778, is remarkable for its integration of geography, geology, climate, vegetation, and anthropology. Forster recognized that plant form and function reflected the local climates in which they grew; this insight was to play a major role in Humboldt's thinking.

Joseph-Louis **Gay-Lussac** (1778–1850) (pp. 79n, 102, 103, 105, 108, 111, 113, 116, 117, 119, 130, 136, 146)

Gay-Lussac was one of Humboldt's closest friends and scientific associates during his tenure in Paris; they shared lodgings for several years. He was a chemist and physicist, associated at various times with the École Polytechnique, the Sorbonne, and the Jardin des Plantes in Paris. Best known for his work on the properties of gases (he discovered both Charles' law and the law of combining volumes), he discovered boron, confirmed that chlorine and iodine were elements, and pioneered volumetric analysis in chemistry. In 1804, he made balloon ascents to 4,000 and 7,000 meters to study changes in air temperature and pressure at different altitudes. Gay-Lussac assisted Humboldt in preparing his volume on physical and chemical properties of the atmosphere following the South American trip, and in 1805 they traveled together to Italy, where they met von Buch for a viewing of an eruption of Vesuvius. During this trip, Humboldt and Gay-Lussac crossed the Alps on foot, measuring magnetic intensity and analyzing gases, which led to their discovery that the composition of the atmosphere does not change with altitude.

Louis **Godin** (1704–60) (p. 106)

Louis Godin, an astronomer and physicist, was the third partner (with Bouguer and La Condamine) in the great geodetic expedition to Peru in 1735. He first conceived the notion for an equatorial geodetic survey and was at the outset its nominal leader. However, frictions developed soon after departure from France, and by the time the expedition reached Peru the three scientists were fighting "like cats and dogs," according to the expedition's surgeon. Godin made the crucial decision to conduct the survey in the Andes rather than coastal Peru, setting the stage for the adventures and disasters that followed. Godin lived a profligate lifestyle, and the great debts he incurred while in Peru led to his detention there, forcing him to take a professorship in Lima. He was finally allowed to return to France in 1751, where he learned that he had been expelled from the French Academy for serving a foreign institution (his letter from Lima explaining his predicament was aboard a ship taken by pirates). Don Jorge Juan and Antonio de Ulloa arranged for him to assume the directorship of the Spanish Naval Academy in Cádiz, where he served until his death in 1760. He never published a narrative of the expedition, and so La Condamine took the greater credit.

Johann Wolfgang von **Goethe** (1749–1832)

Goethe, like Humboldt, eludes a concise biography. He was a central figure in the transition between the dry classicism of the Enlightenment and the personal, intuitive mindset of the Romantic period. Best known for his literary works such as The Sorrows of Young Werther and his monumental poetic drama, Faust, Goethe made direct and indirect contributions to art, philosophy, music, politics, and the sciences. Goethe's scientific works include studies of the physics of color, human anatomy, animal and plant morphology, and plant development. His botanical studies may represent his most lasting contributions to science. Humboldt first met Goethe, a friend of his brother Wilhelm, in 1794, and they became lifelong friends. Humboldt commented that spending time with Goethe was "like being equipped with new organs," and acknowledged Goethe as one of his greatest influences. For his part, Goethe described Humboldt as "a living fountain, whence flow many streams, yielding to all comers a quickening and refreshing draught." Goethe stimulated Humboldt's interest in plant morphology, which became an underpinning of his ideas on plant geography. The influence was mutual; Goethe acknowledged Humboldt's role in shaping his morphological ideas. Goethe's view that landscape had an overriding influence on culture stimulated Humboldt's emphasis on vegetation as a cultural determinant. Humboldt dedicated the German edition of the Essay to Goethe, who in return sketched the comparative profile of Europe and South America.

Thaddeus **Haenke** (1761–1817)   (pp. 80, 86, 90)

Haenke, a Czech native, received his botanical education in Prague and Vienna. His reputation as a botanist led to an invitation from Spain's Carlos IV to accompany Malaspina's global expedition in 1789. He missed the boat, however, arriving in Cádiz a few hours after Malaspina set sail. He found berth with a merchant ship, which sank near Río de la Plata. Haenke finally caught up with Malaspina in Chile, after traveling overland from La Plata. He collected plants with the expedition in South America, Mexico, the Pacific coast of North America, the Philippines, and various Pacific islands. He is credited with the first botanical description of the coast redwood, *Sequoia sempervirens*, which he saw near Monterey. He was also the first botanist to see the giant water lily, *Victoria*. Haenke collected some 15,000 plant specimens on the voyage. Those he sent to Madrid were neglected and ultimately lost, possibly owing to professional jealousies. The duplicate specimens sent to Prague are still intact. He never returned to Europe, choosing to settle near Cochabamba, Bolivia, where he did botanical and geological explorations, developed a silver mine, and established a botanical garden.

Albrecht von Haller (1708–77) (p. 96)

Haller, a Swiss naturalist, worked variously in Bern, London, and Göttingen. Best known for his studies of vertebrate anatomy and physiology, he also did extensive botanical work, in part related to his duties as a physician and medical professor. He conducted widespread botanical studies in Switzerland, which Humboldt alludes to. He also wrote a number of epic poems, plays, and political discourses. Haller wrote a popular account of electricity that inspired Benjamin Franklin's important electrical experiments.

Samuel Hearne (1745–92) (p. 143)

After service in the British Royal Navy, Hearne was employed as a ship's mate and quartermaster by the Hudson's Bay Company. Assigned to explore the region west of Hudson Bay, he made three forays into what is now northwest Canada. His third expedition (1770–72), guided by a Chipewyan chief, Matonabbee, was the most successful; they traversed much of the subarctic and arctic regions of the Northwest Territories, reaching the Coppermine River and following it to its mouth in the Arctic Ocean. Hearne made extensive observations of flora, fauna, and native customs during his journeys.

Francisco Hernández (1514–78) (p. 92)

In 1570, Philip II selected a court physician, Francisco Hernández, to serve as *protomédico* (medical examiner) in the first government-sponsored natural history expedition to the New World. In his seven years in the Valley of Mexico, Hernández described plants and interviewed native healers and physicians, amassing a large store of ethnobotanical information only one to two generations after the Conquest. His work had an enormous influence on botanists, physicians, and naturalists in Europe for the next two centuries. Among those who referenced his observations were Galileo, Aldrovandi, John Ray, Linnaeus, and Erasmus Darwin. Humboldt refers to the "gardens in Iztapalapan, whose remains Hernandez was able to glimpse" (p. 92). These were among several botanical gardens developed by the Mexica (Aztec) culture in the Valley of Mexico, for aesthetic and medicinal purposes.

Nikolaus Thomas Host (1761–1834)   (p. 94)

A native of Croatia, Nikolaus Host studied medicine in Vienna and served for many years as personal physician to Holy Roman Emperor Francis I, who shared Host's interests in natural history. Host was an avid botanist, collecting plants from various parts of the Austro-Hungarian Empire and describing them in a series of monographs and floras. He was the first Director of the Garden for Austrian Plants at Belvedere Palace.

Wilhelm von Humboldt (1767–1835)

Alexander von Humboldt's older brother Wilhelm was a prominent political philosopher, linguist, and diplomat. He and Alexander were very close, though he did not share Alexander's passion for science, natural history, and overseas travel. A friend of Schiller and Goethe, Wilhelm von Humboldt espoused a liberal political philosophy that was ahead of its time for the Germanic states. After founding the University of Berlin (now Humboldt University), he served as a Prussian diplomat in Rome, London, and Vienna, where he played a pivotal role in the Austrian/Prussian/Russian alliance against Napoleon. Wilhelm retired to the Humboldt estate near Berlin in 1819 owing to the increasingly reactionary nature of the Prussian government. He spent the rest of his life in scholarly pursuits, particularly philology and linguistics.

Nikolaus Joseph von Jacquin (1727–1817) and Joseph Franz von Jacquin (1766–1839)   (p. 94)

It is not clear which Jacquin is referred to by Humboldt in the *Essay*. Nikolaus von Jacquin, a native of the Netherlands, had a long career spanning several disciplines, including botany, medicine, chemistry, and mineralogy. He studied medicine and botany at Louvain and Paris, and from1755 to 1759 was part of an expedition to the West Indies commissioned by Francis I, the Holy Roman Emperor of the time. Jacquin collected plants in Martinique, Jamaica, and other parts of the West Indies for the gardens and herbaria of the Schönbrunn Palace in Vienna. He also collected extensively in the Tyrolean Alps and published a flora of Austria following his appointment as director of the Vienna Botanical Gardens in 1768. Nikolaus's son, Joseph von Jacquin, also served as director of the Botanical Garden and taught chemistry and botany at the University of Vienna. The younger Jacquin was a close friend of Nikolaus Host and accompanied him on plant-collecting trips throughout central Europe.

Thomas Jefferson (1743–1826)

Thomas Jefferson is best known as author of the Declaration of Independence and third president of the United States. He had a lifelong interest in science and natural history, and his political views were heavily influenced by his scientific philosophy and experience. Although he made diverse direct contributions to natural history, geography, paleontology, meteorology, and horticulture, his political life diverted him through his lifetime ("Science is my passion, politics is my duty"). Jefferson's most important contributions were vicarious, including his support of the American Philosophical Society, development of the U.S. Coast Survey, foundation of the U.S. Military Academy at West Point (conceived by Jefferson for training of military engineers), and sponsorship of explorations of the western United States. The most notable of these, of course, was that of Meriwether Lewis and William Clark up the Missouri River and down the Columbia to the Pacific coast, which was already underway when Humboldt visited Jefferson in 1805. Jefferson also sponsored other explorations in the west, including the Ouachita River headwaters (led by William Dunbar and George Hunter), the Red River region (led by Thomas Freeman and Peter Custis, but turned back by Spanish troops), the Mississippi River headwaters (Zebulon Pike), and the Arkansas River headwaters (also Pike; interrupted by the incarceration of Pike's party by the Spanish government in Santa Fe). Humboldt visited Jefferson in Washington, their discussions centering on natural history and geography. They corresponded for the rest of Jefferson's life.

Don Jorge Juan y Santacilia (1713–73) (pp. 83, 105, 136, 137, 142)

Don Jorge Juan, a Spanish naval officer, was assigned by the Spanish government to accompany and keep an eye on La Condamine's survey party in South America. Juan proved to be a capable physicist and mathematician for the expedition, and his efforts in barometric determination of altitude were invaluable. He and another officer, Antonio de Ulloa, were also charged with preparing a report for King Philip V on conditions in Peru. Their report, unsparing in its description of bureaucratic and clerical corruption and abuse of Indian laborers, had little influence; Philip died just after Juan's return to Spain.

Antoine Laurent de Jussieu (1748–1836) (p. 59)

Antoine Laurent de Jussieu spent his scientific career at the Jardin du Roi (now Jardin des Plantes) and the Muséum national d'Histoire naturelle in Paris. He was demonstrator and, later, professor of botany. Jussieu adopted Linnaean nomenclature but substituted an alternative, "natural" classification of plants based on a wider and less rigid set of characters than Linnaeus advocated. His taxonomic revisions were influential in the development of evolutionary theory in the nineteenth century, and his higher-level classification scheme was retained until the late twentieth-century advent of cladistics. Many families he named are still widely used. Jussieu was from a distinguished family of botanists; his uncles included Antoine (professor of botany at the Jardin du Roi), Bernard (demonstrator of botany at the Jardin), and Joseph (a botanist and physician who went to South America in 1735 with La Condamine, was repeatedly detained by local authorities because of his valuable medical skills, and could not return to France until 1771).

Carl Sigismund **Kunth** (1788–1850)

Kunth, the nephew of Humboldt's boyhood tutor, was working as a clerk in Berlin when he met Humboldt and Willdenow, both of whom encouraged him to study botany. Kunth worked with Willdenow in Berlin until the latter's death in 1812, whereupon Humboldt invited him to Paris to perform the exacting systematic work on the vast plant collections of Humboldt and Bonpland. He took up the work that Bonpland had left unfinished, spending seventeen years working in Paris and writing a series of monographs on the plant collections. Kunth returned to Berlin in 1829, spending the rest of his life there at the university and botanical garden.

Charles-Marie de **La Condamine** (1701–74)   (pp. 83, 84, 89, 93n, 105, 105n, 106, 136, 137, 137n, 146)

La Condamine, a French mathematician and physicist, was one of three leaders of the expedition that now bears his name owing to his vivid and popular narrative. Together with two other members of the Royal Academy of Sciences, Pierre Bouguer and Louis Godin, he set sail for South America in 1735 to measure the length of a meridional degree near the equator, which would indicate whether the globe was an oblate (flattened) or prolate (elongated) spheroid. The expedition faced numerous hardships, including squabbles among the leaders, adverse terrain and weather, and tensions with the local populace. La Condamine, Bouguer, Juan, and others narrowly escaped with their lives when a riot erupted at the bullring in Cuenca. Their surgeon was killed and Bouguer stabbed in the melee. Despite the difficulties, the expedition was a success, with signal advances in geography, physics, and natural history. La Condamine capped his adventure by traveling from the east slope of the Andes to the mouth of the Amazon aboard rafts and dugouts. His botanical contributions included bringing knowledge of rubber trees, quinine (cinchona) trees, and barbasco, the natural source of rotenone, to European attention. All three were used by native populations in South America, but they were largely unknown in Europe. After the expedition he was a tireless advocate for a uniform system of measurement and for inoculation against smallpox. He died following hernia surgery, during which he asked the surgeon to work slowly so that he could observe and prepare a report to the Academy.

Philippe-Isidore Picot de **Lapeyrouse** (1744–1818) (p. 94)

Lapeyrouse was a prominent naturalist and politician of the Languedoc region of southern France. Born to a wealthy family of the region, he was trained as a law-yer and served as an administrator and magistrate in Toulouse until the French Revolution. He had strong republican sympathies and served in the revolutionary governments of the region until his arrest and imprisonment during the Terror. La-peyrouse had a passion for natural history, and pursued extensive field studies of flora, fauna, fossils, and minerals of Languedoc and the adjacent Pyrenees. He pub-lished numerous works on these topics throughout his adult lifetime. Lapeyrouse served as mayor of Toulouse during the Napoleonic era and was reappointed during the Hundred Days of Napoleon's return in 1815. He was forced into hiding for sev-eral months after the Bourbon restoration to avoid imprisonment or execution by the Verdets, vengeful royalist gangs intent on cleansing Languedoc of all traces of republicanism.

Pierre-Simon **Laplace** (1749–1827) (pp. 62, 83, 103, 108, 114, 115, 136, 137, 147)

Laplace was a giant among giants in the science of his time, making major contribu-tions to mathematics, physics, astronomy, demography, and statistics. His talents were recognized early by d'Alembert, one of the leaders of the French Enlighten-ment, who arranged for a teaching position at the École Militaire in 1768. Follow-ing the Revolution, he was appointed director of both the Paris Observatory and the new Bureau des longitudes, though he was not entirely suited to the administrative requirements of the former nor the practical expectations of the latter. Napoleon appointed him Minister of the Interior in 1799, but removed him a few weeks later "because he brought the spirit of the infinitely small into the government." Hum-boldt's thinking about earth history was strongly shaped by Laplace's conception of the formation of the earth and planetary system. Laplace's five-volume masterwork, *Traité de mécanique céleste*, was preceded by a lengthy and exhaustive introductory vol-ume for the general reader, *Exposition du système du monde*. Asked by Napoleon why he never mentioned God in his book, Laplace is said to have responded, "I have no need of that hypothesis."

Alessandro **Malaspina** (1754–1810) (pp. 80, 83)

Malaspina, a Tuscan native, became an officer in the Spanish Navy in the 1770s. He led an expedition to survey Spanish possessions of the Americas and Asia from 1789 to 1794, during which they explored the coasts of the Alaska panhandle, Vancouver Island, Puget Sound, California, and Mexico, as well as Australia and the islands of the western Pacific. After the expedition, he was outspoken in his recommendations for political and economic reform in the Spanish colonies. He was arrested in 1795 and imprisoned in Coruña until an international amnesty campaign resulted in his release late in 1802. He was exiled from Spain and spent his remaining years in Italy.

William **Marsden** (1754–1836) (pp. 98, 138)

Born and educated in Dublin, Marsden served the East India Company in Sumatra, where he mastered the language and made close observations of the culture and geography. He returned to England and published his now-classic *History of Sumatra* in 1783. He served the British Admiralty as a high-level bureaucrat and was widely known as an orientalist and scholar in his time.

Pierre François André **Méchain** (1744–1804) (pp. 118, 141)

Despite his considerable native talents in mathematics, Méchain, the son of a plasterer, was financially unable to pursue studies in engineering, so he took up astronomy as a hobby while earning money as a tutor. He was forced to sell his telescopes when financial calamity struck his family, but the buyer turned out to be Jérôme Lalande, a prominent and well-connected astronomer. Lalande secured him a position as a naval cartographer at Versailles, where he continued his astronomical work and discovered a number of comets and other celestial objects. In the 1790s, Méchain was, with Delambre, the leader of the effort to measure the meridional distance between Dunkirk and Barcelona. Méchain measured the southern transect, surviving arrest in France, imprisonment in Spain, and various injuries and illnesses sustained during the field work. He became director of the Paris Observatory in post-revolutionary France. Méchain was a perfectionist, driving himself relentlessly. Obsessed with errors in his portion of the meridional survey, he returned to Spain in 1803 to repeat his measurements. After running transects through a series of coastal marshes and ricefields, he contracted malaria and succumbed a few weeks later.

Daniel **Melanderhielm** (1726–1810) (p. 118)

Melanderhielm was a prominent Swedish astronomer and mathematician, based at the University of Uppsala.

José Mariano **Mociño** (ca. 1760–1820) (p. 80)

José Mociño, a native of Mexico, was a broadly trained naturalist who from 1791 to 1804 worked closely with Martin Sessé in surveys of the flora and natural history of Spanish territories, including California, Mexico, Central America, Cuba, and Puerto Rico. This monumental survey, the Royal Botanical Expedition, led to thousands of pressed plants and thousands of drawings and paintings by Atanasio Echeverria, who was part of the expedition. Mociño, Sessé, and Echeverria traveled to Spain in 1803 to prepare the publications from the expedition, but they were soon caught up in the political turmoil of the Napoleonic era. Mociño was accused of collaborating with the French during Napoleon's occupation of Spain, and he was put in a chain gang for a time. He fled Spain with his drawings and manuscripts, but the latter were lost when his cart was seized by the French military. His drawings were eventually copied and preserved by Augustin de Candolle. Results of the expedition (*Plantae Novae Hispaniae* and *Flora Mexicana*) were published in 1887, long after the deaths of all the participants.

Carlos **Montúfar** y Larrea (1780–1816) (p. 146)

Carlos Montúfar was a son of the Marqués de Selvalegre, who was provincial governor in Quito during Humboldt's visit there in 1802. He accompanied Humboldt and Bonpland in their explorations of the Andes, including the Chimborazo ascent, and traveled with them to Mexico, Cuba, the United States, and France. In France, Montúfar befriended Simón Bolivar, whom Humboldt had met in Caracas. Montúfar's political views were undoubtedly influenced by Humboldt and Bolivar, by his visit to Philadelphia and Washington (where he met Jefferson), and by his extended stay in Paris. He spent time in Spain and joined the resistance forces against Napoleon's invasion during the Peninsular War, attaining the rank of lieutenant colonel. Upon his return to South America in 1812, Montúfar joined the revolutionary forces, fought in several battles, was betrayed and imprisoned, but escaped in time to join Bolivar's 1814 capture of Bogotá. He was captured and executed by Morillo's forces near Popayán.

Don José Celestino **Mutis** (1732–1808) (pp. 80, 89, 108)

Mutis, a Spanish native, was trained as a physician and botanist. In 1760 he emigrated to South America, where he became the continent's foremost naturalist and scientist of his time. He founded an observatory in Bogotá, surveyed the local flora, and corresponded extensively with Linnaeus. He envisioned a grand survey of New Granada (Colombia), which was eventually approved and funded by Carlos III in

1783. The survey was still underway when Humboldt and Bonpland visited him in 1802, and it in fact remained uncompleted after his death. He amassed an herbarium of 24,000 plant specimens, commissioned detailed drawings of 5,000 plant species, assembled a botanical library second only to that of Sir Joseph Banks, and collected innumerable zoological specimens. Publication of his monumental *Flora de la Real Expedición Botánica del Nuevo Reino de Granada* remained incomplete until the late twentieth century. Mutis conducted extensive studies of the various species of cinchona (quinine-bark).

Peter Simon **Pallas** (1741–1811) (pp. 94, 99)

Pallas, a German-born naturalist, was appointed professor at the St. Petersburg Academy of Science by Catherine II in 1767. He led two major expeditions, the first to Siberia, which took him to the Caspian Sea, the Amur drainage, the Altai Mountains, and Lake Baikal, and the second to the Black Sea and Crimean Peninsula.

José Antonio **Pavón** y Jiménez (1754–1844) (pp. 92, 93)

Pavón was the junior partner in the Ruiz/Pavón/Dombey botanical expedition to Peru and Chile (1778–88). He got along well with both Ruiz and Dombey, but seems to have had less energy and ambition than the other two. He was not up to the task of completing the *Flora* following Ruiz's death.

Louis-Marie Aubert **du Petit-Thouars** (1758–1831) (p. 98)

Born into an aristocratic family, Louis-Marie du Petit-Thouars served as a naval officer, was imprisoned during the French Revolution, and then exiled to Madagascar. An accomplished botanical scholar, du Petit-Thouars spent his years in exile cataloging the flora of Madagascar, Mauritius, and the Île Bourbon (now La Réunion). He described more than a hundred new species of orchids, including *Angraecum sesquipedale*, later made famous by Darwin's description of a hitherto unknown moth based solely on its floral anatomy.

Gaspard Clair François Marie Riche de **Prony** (1755–1839) (p. 62)

Prony, who assisted Humboldt in his barometric calculations of altitude, was a prominent French engineer and mathematician. He directed the effort at assembling the *Cadastre*, the first systematic compilation of precise logarithmic and trigonometric tables, which required the efforts of seventy to eighty "human calculators" working for nearly a decade.

Louis-François **Ramond**, Baron de Carbonnières (1755–1827)

(pp. 62, 86, 94n, 96, 132, 133, 141)

Ramond, a French botanist, did extensive surveys of the floras of southern France. He is best known for his 1789 monograph on the natural history of the Pyrenees. Ramond's discussion of elevational patterns in vegetation and flora influenced Humboldt's early thinking on the subject. A constitutionalist/royalist, Ramond served in the Assembly during the revolutionary period but fled to the Pyrenees as the Terror began. After eighteen months in hiding (during which he did extensive botanical and geological studies), he was discovered, arrested, and imprisoned early in 1794. He narrowly escaped the guillotine.

Andrés Manuel **del Río** (1764–1849)

(p. 125)

A native of Spain, Andrés del Río studied mineralogy with Werner at the School of Mines in Freiberg, where he befriended Humboldt, a fellow student there. He then traveled to Paris to study chemistry with Lavoisier, but his studies were cut short by Lavoisier's arrest and execution during the Terror. He emigrated to Mexico City in 1794 to assume a chair in chemistry and mineralogy at the new Royal College of Mines. Del Río had a long and productive career in Mexico, researching widely in mineralogy, petrology, and stratigraphy. He developed a high-quality iron-works, which was destroyed by Royalist forces during the Mexican war of independence. Humboldt visited him in 1803, which ultimately led to a misunderstanding. Del Río had discovered a new element, vanadium, but a French chemist to whom Humboldt sent samples concluded the material was chromium. Del Río was vindicated nearly thirty years later. He was exempted from the series of post-revolution expulsions of Spanish citizens, but in 1829 he went into self-imposed exile in Philadelphia to protest the expulsion of Spanish colleagues. He returned to Mexico in 1834 after the fall of the Guerrera and Bustamante dictatorships.

Hipólito **Ruiz** López (1754–1815)

(pp. 92, 93)

Ruiz was part of a botanical expedition to Peru, sponsored by Carlos III, that also included José Pavón, the French botanist Joseph Dombey (who provided on-the-fly botanical training to the two Spaniards), and two artists. They embarked for Peru in 1778 on a ship full of fortune-seekers, who viewed the botanists as imbeciles for making the long and perilous voyage for the purpose of collecting plants. Originally intended as a four-year survey, a variety of factors (the Andean rebellion of Tupac Amaru II, a two-year botanical excursion to Chile, and additional orders from the government) delayed return of Ruiz and Pavón until 1788. Conflicts arose periodically between Ruiz and Dombey, who returned to Europe in 1785 owing to health reasons. (Dombey met a tragic end, being captured by the British in 1794 while he was en route to the United States to advocate for the new metric system. He died in a prison on Montserrat.) Ruiz seems to have been scrappy; after the expedition he was embroiled in disputes with Cavanilles, Mutis, Pavón, and others. The expedition was overall successful, although many specimens and records were lost in a fire in a remote Andean village and in a shipwreck off the coast of Portugal. Thousands of plants were illustrated and described, and three volumes of the Flora of Peru were published between 1798 and 1802. A fourth volume was not published until the mid-twentieth century. The remaining eight volumes were never completed because of cost overruns for the first three volumes, the disruptions of the Peninsular War, and Ruiz's death. Humboldt and Bonpland met Ruiz and Pavón and examined their plant collections in Madrid just before departing for South America.

Horace Bénédict de **Saussure** (1740–99)   (pp. 95, 96, 102, 105, 109, 110, 112, 113, 115, 132, 136, 139, 140, 146)

Saussure, a native of Geneva, was fascinated by mountains and devoted his life to studying botany, geology, meteorology, and physics of the Alps. He explored the Alps extensively and sponsored the first ascent of Mont-Blanc in 1786. He climbed to the summit himself the following year. His scientific contributions are diverse. He coined the term geology, did pioneering work in stratigraphy, recognized the geomorphic influences of glaciers, and demonstrated that plant chemistry and morphology are influenced by mineral composition of soils. Saussure invented a variety of instruments, including devices for measuring wind speed, magnetic fields, the blueness of the sky, the clarity of the atmosphere, and temperature in unusual places (deep waters, glaciers, soils). He developed the first electrometer for measuring electrical potential, and invented the hair hygrometer, a device for measuring atmospheric humidity based on absorption of water by human hair. Humboldt took several instruments of Saussure's design on his voyage, and his fascination with mountains as natural laboratories was rooted in Saussure's work. Humboldt met Saussure in 1796 when he passed through Switzerland on his way back from Italy, and was quick to apply Saussure's ideas on the movement of glaciers as he journeyed through the Alps.

Franz von Paula **Schranck** (1747–1835)   (p. 94)

Schranck, a German Jesuit priest, developed interests in botany and entomology as a missionary in Brazil. After returning to Europe, he spent much of the rest of his life in Bavaria, where he taught mathematics and physics. He wrote several botanical monographs for Bavaria and adjacent Austria. In his last years he served as director of the Botanical Gardens in Munich.

Giovanni Antonio **Scopoli** (1723–88) (pp. 68, 87)

Scopoli was a physician and naturalist who spent his life in Italy, Austria, and Slovenia. His interests were diverse; he published works on floras and faunas of various regions of the Alps, a treatise on mercury poisoning in miners, and a comprehensive dictionary of chemistry. The drug scopolamine was named for him.

Martin de **Sessé** y Lacasta (1751–1808) (p. 80)

A native of Spain, Sessé served as a military physician in Mexico City, where he taught botany at the Pontifical University and founded the Royal Botanical Garden in 1788. Sessé petitioned the viceroy to support an ambitious survey of the natural history of the colonies of New Spain, and Carlos III authorized the Royal Botanical Expedition. This expedition, led by Sessé and the Mexican naturalist José Mociño, lasted from 1789 to 1803 and covered a vast territory, ranging from Vancouver Island to Nicaragua, and from Mexico to the Spanish islands of the Caribbean. Sessé and Mociño were summoned to Spain in 1803, where they planned to prepare the results of the expedition (including 8,000 plant specimens and hundreds of birds and fish) for publication. Sessé's untimely death in 1808, and the political and economic instability during and following the Napoleonic era, delayed publication of the results of the expedition until 1887. Humboldt and Bonpland viewed many of the Mexican plant collections and illustrations made by Sessé and Mociño at the Museum of Natural History in Madrid in 1799.

George Augustus William (pp. 105, 138, 139, 140)
Shuckburgh-Evelyn (1751–1804)

Shuckburgh, born to an English baronetage, was a prominent mathematician and astronomer. He made many observations of the lunar surface from his personal observatory at his estate. He served in the House of Commons for many years.

Johann Volkmar **Sickler** (1742–1820) (p. 62)

Sickler was a pioneer in the field of pomology, the cultivation of fruit-tree orchards. He imported a number of new varieties of fruit trees to Europe, wrote several books on pomology, launched a magazine, Der Teutsche Obstgaertner (The Teutonic Fruit Gardener), and established orchards and nurseries in Thuringia. Sickler's first nursery, which included 8,000 grafted trees, was destroyed in 1806 when a corps of Napoleon's army encamped in the orchard following the Battle of Jena. He developed a new nursery, which was destroyed by Cossacks following the Battle of Leipzig in 1814. Not easily discouraged, he established a third nursery in 1817.

Kaspar Maria von **Sternberg** (1761–1838)   (pp. 98, 140, 141)

Count Sternberg was a Bohemian cleric and naturalist who did pioneering work in paleobotany, particularly the fossil plants and stratigraphy of the Carboniferous. In an official visit to Paris for Napoleon's 1805 coronation, he visited Humboldt and others in the local scientific community, which led to his interest in paleobotany and geology. He also did floristic and systematic work in the mountains of central Europe, corresponded with Goethe on botany and minerals, and founded the Bohemian National Museum in Prague.

Louis-Nicolas **Vauquelin** (1763–1829)   (p. 88)

Vauquelin, a French chemist and pharmacist, assayed many soil and mineral samples that Humboldt collected in his voyage to South America. Together with Antoine-François de Fourcroy, his predecessor as professor of chemistry at the Université de Paris, he analyzed guano samples Humboldt collected in Peru. They observed that the samples had higher levels of nitrogen and phosphorus than any known fertilizer. This ultimately led to the nineteenth-century "guano rush" on the Pacific coast of South America. Vauquelin is credited with the discovery of several elements, including beryllium, chromium, osmium, and lithium, and with the first isolation of an amino acid, asparagine. He also did important work in organic chemistry, particularly of natural plant compounds (pectin, malic acid), and did pioneering studies of the chemical composition of the brain. Vauquelin's liberal political views led to dismissal from his academic posts in 1822.

Dominique **Villars** (1745–1814)   (pp. 94, 95, 97)

Villars was a physician and botanist based in Grenoble. He developed a botanical garden in Grenoble and published a flora of the Dauphiné region of southeastern France. As a physician, Villars worked extensively with the rural poor in remote regions of the Alps. He became a medical reformer, arguing in particular for empirical testing of the medicinal properties of plants used in folk medicine.

Constantine François Chaseboeuf Boisgirais,   (pp. 99n, 143)
Comte de **Volney** (1757–1820)

Volney, an important French political philosopher, is best known for his 1791 book, *The Ruins: or a Survey on the Revolutions of Empires*, a sweeping survey of ancient history, religious skepticism, and political science which influenced people from Percy Bysshe Shelley to Abraham Lincoln. His opposition to Robespierre led to his imprisonment during the Terror. He recognized Napoleon's military and organizational talents, and fostered his reinstatement in the French army following the Terror. Volney was a friend of Franklin's and an admirer of the new American republic, and in 1795 he emigrated to Philadelphia with intention to settle permanently in the United States. He was elected to the American Philosophical Society, but was suspect in the view of the Adams administration owing to his liberal political views, French citizenship, and association with Jefferson and other republicans. Volney was a primary target of the notorious Alien and Sedition Acts, passed in 1798 to suppress dissent and expel foreigners with questionable political leanings. He returned to France in 1799. During Volney's sojourn in the United States, he researched and wrote a volume on the climate and soil of the United States (published in 1804), which is the work cited by Humboldt.

Alessandro **Volta** (1745–1827) (pp. 112, 116, 148)

Volta was an Italian physicist who did pioneering research on electricity and electrochemistry. He invented the first battery in 1795, primarily as a means of settling a dispute with Galvani over the existence of "animal electricity." Humboldt was fascinated by the dispute and in 1797 published a monograph summarizing hundreds of experiments he conducted on the subject, many involving painful experiments on his own muscles. Humboldt visited Volta at his Lake Como estate in 1796, and dropped in again with Gay-Lussac during their 1805 journey to Italy.

Abraham Gottlob **Werner** (1749–1817) (p. 123)

Werner, a Saxon geologist, is best known as the leading exponent of "Neptunism," the idea that all rocks—including crystalline and massive rocks (e.g., granite, basalt) as well as sedimentary rocks (limestones, shales) derived from water-dominated processes. He published little in his lifetime but had a far-reaching influence through his students, whom he attracted from all over Europe to the Academy of Mines in Freiberg. Humboldt, one of his students, was initially an enthusiast for Neptunism, but became an apostate after careful consideration of granites and volcanic activity in the New World. Humboldt repudiates Neptunism in the *Essay*.

Carl Ludwig **Willdenow** (1765–1812) (p. 62)

Humboldt first met Willdenow in 1788, shortly after Willdenow had published a flora of Humboldt's native Berlin region. Willdenow was interested in the geography of plants, and he stimulated Humboldt's passion for botany. Indeed, Willdenow was second only to George Forster in his influence on Humboldt's early thinking on plant geography. Willdenow ran an apothecary shop in Berlin until 1798, when he was appointed professor of natural history at the Berlin Medical and Surgical College. In 1801 he was appointed director of the Berlin Botanical Garden. He published the first edition of his *Principles of Botany* in 1792. The book was later translated into English and was the first to lay out a systematic set of principles of the geography of plants. These served as a springboard for Humboldt's synthesis. After Humboldt's return from the Americas and it had become clear that Bonpland was not up to the task of systematic description of the plant collections, Humboldt enlisted Willdenow to do the detailed botanical work. Willdenow died before he was able to accomplish much, so Humboldt recruited Kunth to finish the work.

Helen Maria **Williams** (1761–1827)

Helen Maria Williams published several volumes of poetry and a novel in London during the 1780s. The poems, often on anti-war, anti-imperialist, and anti-slavery themes, were acclaimed by Robert Burns and William Wordsworth, among others. Intrigued by the French Revolution, Williams traveled to France in 1790 to witness events firsthand. She never returned to England. Her eight-volume *Letters from France* is regarded as one of the best eyewitness accounts of the various phases of the Revolution and its aftermath. As a close friend of many Girondins, Williams was arrested and imprisoned during the Terror, and many of her friends were guillotined. Initially supportive of Napoleon Bonaparte, she became an increasingly vocal critic of his belligerent foreign policy. Her public writing career was interrupted in 1803 by Napoleon's censors and did not resume until Napoleon's ouster in 1815. She met Humboldt in Paris shortly after his return from the Americas, and they developed a lifelong friendship. He was impressed by "her eloquent pen" and asked her to translate his seven-volume *Personal Narrative* and two-volume *Researches* from French into English. This provided much-needed income during a period of financial stress, though she was understandably pleased to get the monumental task completed. Williams's translations were the first English versions of any of Humboldt's works and were read widely in England and the United States. She spent her final years working for political rights and freedoms in France, though she continued to write poetry throughout her lifetime.

Franz Xaver von **Wulfen** (1728–1805)   (p. 94)

The son of an Austrian field-marshal, Wulfen became a Jesuit priest and spent his life teaching mathematics and physics in Vienna and other cities of Austria. Wulfen was a passionate botanist and mountaineer. He explored the eastern Alps extensively, collecting plants and minerals. Wulfen published his floristic studies of the Carinthian region of Austria and Slovenia, and was a correspondent of Linnaeus and other contemporary botanists. The mineral wulfenite was named in his honor.

Eberhardt August Wilhelm von **Zimmermann** (1743–1815) (p. 132)

Zimmermann spent his career as a professor of mathematics and physics at the University of Braunschweig, where he mentored Carl Friedrich Gauss. His most lasting scholarly contributions were in geography and natural history. He was one of the pioneers of zoogeography. His works included the 1777 *Specimen Zoologicae geographicae, quadrupedum domicilia et migrationes sistens*, the first systematic attempt at mammalian zoogeography on a global scale. Zimmermann argued that the traditional biblical notion that all animal species had dispersed worldwide from Mt. Ararat was insupportable. His arguments paved the way for the biogeographic syntheses of Lyell, Hooker, Darwin, Wallace, and others. Zimmermann also recognized the role of climate in influencing global distributions of animals.

Bibliographical Essay and Bibliography

Stephen T. Jackson

Bibliographical Essay

HUMBOLDT'S *ESSAY*

Humboldt's *Essay* was published in Paris in 1807 and simultaneously in Tübingen as *Ideen zu einer Geographie der Pflanzen*. Although it is occasionally referred to as an 1805 work, and some copies bear an 1805 imprint, there is no evidence of actual publication in that year. The confusion apparently stems from Humboldt's presenting the work in 1805 to the *Classe des Sciences Physiques et Mathématiques* (formerly the *Académie Royale des Sciences*) in Paris. Humboldt also referred to it on occasion as an 1805 work. The *Essay* was printed in a very limited edition and was never reprinted in Humboldt's lifetime. It has been reprinted in facsimile five times in the twentieth century:

1959 by the Society for the Bibliography of Natural History (London). This version lacks the *Tableau physique* and related essay. Copies are still available for purchase from the publisher.

1963 by Wissenschaftliche Buchgesellschaft (Darmstadt). This is a facsimile of the German *Ideen* and contains an accompanying essay by Adolf Meyer-Abich, a black-and-white print of the German-language version of the Chimborazo profile, and a print of the French version of Goethe's profile sketch of the Andes and Alps.

1965 by the Institut Panaméricain de Géographie et d'Histoire (México City). This folio-size facsimile includes a full-size (94 x 64 cm) color reproduction of the *Tableau physique*.

1977 by Arno Press (New York) as part of the *History of Ecology* series (edited by Frank N. Egerton). This edition lacks the *Tableau physique*.

1990 by Éditions Européennes Erasme (Nanterre) as part of the *Memoria Americana* Collection, edited by Charles Minguet, Amos Ségala, and Jean-Paul Duviols. This attractive folio edition includes prefatory material in French by Charles Minguet and Jean Théodoridès, selected color plates from the Humboldt/Bonpland/Kunth botanical monographs, a color print of Goethe's

German-language sketch of the Andes and Alps, and a reduced-size (29 x 48 cm) color print of the *Tableau physique.*

An English translation of the first two parts, the "Preface" and "Essay," of the *Essay* (by Francesca Kern and Philippe Janvier) was published in the 2004 compendium, *Foundations of Biogeography* (Lomolino et al., eds.). This printing lacks the *Tableau physique* and related essay. It is a fine translation, though printed in a very small font. A Spanish translation of the *Ideen* was published in Bogotá in 1985. This volume also contains several botanical writings of Caldas, including his "Preface" to the translation of Humboldt's *Essay* that was published in *Seminario de la Nueva Granada* in 1809, a series of undated notes on Humboldt's early draft, and an essay on the origin of cultivated plants in the Andes. It also features a reprint of the "Physical Tableau" essay in Spanish translation.

HUMBOLDT'S OTHER WRITINGS

If you have time to read only one book by or about Alexander von Humboldt, read his *Personal Narrative.* If you do not have a few months to devote to the work, read the recent one-volume translation by Jason Wilson. It is nicely done, short enough to be digested in the course of a couple of weeks, and as a bonus the volume includes informative essays by Wilson and by Malcolm Nicolson. This edition is heavily abridged, taking 3,927 pages in seven volumes from Humboldt's original and distilling them into 297 pages.

Two other English translations of the *Personal Narrative* have been published, both in the nineteenth century. Both are out of print. The earliest translation is by Helen Maria Williams, a native of England who worked closely with Humboldt (who was reasonably fluent in English) in preparing the translation. The Williams version, published in multiple volumes between 1814 and 1829 by Longman, Rees, Orme, Brown, and Green (London), is the closest we have to an "official" translation and is unabridged from the original French version. Thomasina Ross prepared a second translation, published in London by G. Bohn in 1850 and republished in the late nineteenth century as part of Sir John Lubbock's "Hundred Books" series. The Ross translation is abridged, extending through 1,433 pages in three volumes. Its primary deletions from Humboldt's original consist of the technical and scientific digressions and tabular materials. It was intended for a general audience and has an index, which the Williams translation lacks.

I prefer the Williams translation, for three reasons. First, it was prepared in close collaboration with Humboldt and may be considered to be closest to his wishes. Jason Wilson, in his translator's introduction to the 1995 edition, suggests that Williams invoked some poetic license in translating Humboldt's more lyrical passages, but Nigel Leask has pointed out that Humboldt may well have intended for her to do so. Second, the Williams translation contains everything—everything!—Humboldt incorporated into the original French manuscript. All the 100-page scientific digressions are there, all the enumerations of facts and figures, all the statistical tables. It is only in this version that the reader gets a true sense of the breadth and depth of Humboldt's interests and insights. It makes for intimidating reading, but the longueurs can always be skipped or skimmed when necessary. My third, and perhaps most important, reason for preferring Williams is that it is her translation that was read by the young Charles Darwin, along with Alfred Russel Wallace, Charles Lyell, Joseph Dalton Hooker, Asa Gray, Henry David Thoreau, Ralph Waldo Emerson, Washington Irving, Frederic Church, and countless others in England and North America. The original version is, of course, expensive and difficult to obtain. However, it has been reprinted twice, once in 1966 by AMS Press and again in 1972 by Da Capo Press and simultaneously in Amsterdam by Theatrum Orbis Terrarium. These sets were widely purchased by libraries and should be easily found in university and large municipal libraries.

Humboldt's collection of popular essays, *Ansichten der Natur*, published in Tübingen in 1808, was published twice in English in 1850, as *Aspects of Nature*, translated by Elizabeth Leeves Sabine ("Mrs. Sabine." 1807–79) and as *Views of Nature*, translated by Elise C. Otté (1818–1903) and Henry G. Bohn (1796–1884). Both are from the 1849 third edition in German. I find the Otté/Bohn translation a little smoother to read. Both English versions are out of print. This is unfortunate, because *Views of Nature* is one of Humboldt's most important works, one he described as his personal favorite. It consists of seven essays, six of which were drafted during his travels and focus on natural history and science. Humboldt is at his most lyrical in these brief essays, which have some lovely passages. Each essay is accompanied by "illustrations and additions," extended footnotes that comprise nearly 70 percent of the book. All of the essays are worth reading; they provide a good sense of Humboldt's catholic interests. Essays on "steppes and deserts" and "physiognomy of plants" are particularly useful in getting a sense of Humboldt's ecological thinking.

Researches Concerning the Institutions and Monuments of the Ancient Inhabitants of the Americas, with Descriptions of Some of the Most Striking Scenes in the Cordilleras, was published in Helen Maria Williams's English translation in 1814. It was reprinted in Amsterdam in 1972 but is out of print now. This work consists of a series of essays written around an atlas of plates and lithographs. Most are archeological or anthropological (e.g., sculptures, earthworks, and calendars of the Aztecs and Incas; codices archived in European libraries). Others focus on "views"—lithographs of various landmarks and incidents relating to Humboldt's voyages.

Humboldt's crowning work was his five-volume, never-finished *Cosmos: A Sketch of a Physical Description of the Universe*, published originally in German between 1844 and 1859. Humboldt aimed at nothing less than a comprehensive description of all known phenomena in the universe, ranging from celestial bodies and mechanics to the properties of the human mind. Human perceptions and aesthetics interpenetrate the entire work but are most prominent in the astonishing second volume (1847). Here he describes the history of human perceptions of the natural world, including a comprehensive history of scientific thought in all documented cultures and an extended discussion of landscape painting, vegetation form, and botanical gardens as the primary portal for appreciation of nature. Humboldt's unified vision of the sciences and his sense of the interconnectedness of organic and inorganic nature are major themes of *Cosmos*.

All five volumes have been translated into English (and many other languages). Elizabeth Leeves Sabine, wife of Humboldt's British geomagnetic collaborator Edward Sabine, translated the first four volumes, which were published in England by Longman, Brown, Green, and Longman. Copies of the Sabine translation are scarce today. All five volumes were also translated by Elise C. Otté and published in London by Henry G. Bohn and in New York by Harper and Brothers. Copies of the Otté translation are more easily found, and the first two volumes were reprinted in 1997 by Johns Hopkins University Press. Essays by Nicolaas Rupke and Michael Dettelbach accompany these latter reprints. The first two volumes are by far the most rewarding for the modern reader.

A translation of Humboldt's four-volume *Political Essay on the Kingdom of New Spain* by John Black was published in 1811 and was reprinted in 1966 by AMS Press. A heavily abridged version was in print as recently as 1988; it omits the maps, tables, and Humboldt's pioneering statistical graphs. Book 4 ("State of the Agriculture of New Spain") contains extensive discussions of the climate, vegetation, soil, and agriculture of Old Mexico. The *Political Essay on the Island of Cuba* has also been published in multiple English translations, none of which seem to be currently in print. The *Island of Cuba* includes a passionate condemnation of slavery (omitted for political reasons from an unauthorized 1856 translation published in Philadelphia). It also contains descriptions of the landscape of the island, including its soils, climate, and agriculture. The *Island of Cuba* comprises much of the final volume (7) of Humboldt's *Personal Narrative* and hence is available in the reprint editions of that work.

A condensed and edited account of Humboldt's New World voyage was prepared in 1832 by William MacGillivray (1796–1852), a Scottish naturalist and surgeon who also edited John James Audubon's *Ornithological Biography* and the text to Audubon's *Birds of America*. MacGillivray made liberal use of Humboldt's *Personal Narrative* and *Political Essay on the Kingdom of New Spain*, and added biographical information from other sources. It went through several revisions and editions until MacGillivray's death. This volume was highly popular in England and North America, and remains highly readable. To my knowledge it has not been reprinted.

Humboldt kept extensive journals (in French, German, and occasionally Spanish) during his travels, which formed the basis for the *Personal Narrative*. A lovely two-volume set, *The Route of Humboldt: Colombia and Venezuela*, designed and edited by Benjamin Villegas, combines an English translation of the Venezuelan portion of the *Personal Narrative* with excerpts of the journals for Colombia. This set is enriched with photographs and illustrations of the regions Humboldt visited. Unfortunately, the volume contains journal entries only for Columbia; it breaks off near Pasto, Colombia, where Humboldt passed into what is now Ecuador. Thus, Humboldt's notes on his ascent of Chimborazo as well as his remaining travels in Ecuador, Peru, and Mexico remain unavailable in English.

HUMBOLDT'S LIFE

There are several English-language biographies of Humboldt, ranging from nineteenth-century hagiography to twenty-first-century metabiography. By far the best nineteenth-century work is the two-volume, 1873 translation of the Karl Bruhns "biography." It is really three biographies written by three different people (none of them Bruhns), each focusing on a different period and aspect of Humboldt's life. It is still widely consulted. However, the Bruhns volume, like other nineteenth-century German biographies (e.g., Klencke 1852), devotes minimal attention to Humboldt's twenty-two-year sojourn in Paris and largely neglects his extensive French-language publications (including the *Personal Narrative* and the *Essay*). The only other nineteenth-century biography of note is by the American travel writer and literary critic Bayard Taylor. Taylor visited Humboldt in Berlin in 1856 and 1857. The biography, published shortly after Humboldt's death, draws heavily from Humboldt's own writings.

The three most influential twentieth-century biographies are by Helmut de Terra (1955), Lotte Kellner (1963), and Douglas Botting (1973). De Terra's biography is detailed, comprehensive, and personal, which has made it somewhat controversial. Kellner's is comprehensive and thoughtful, with less emphasis on the voyage and more on the scholarly work than the other biographies. Botting's book is also well researched and attractively produced, with numerous maps and figures, many in color. It focuses on Humboldt as scientist and explorer. Gerard Helferich's *Humboldt's Cosmos* is a fine narrative of Humboldt's travels, sandwiched between brief before and after chapters. It is well researched and well written, though the Tenerife and Venezuela portions of the journey are (in my opinion) preferably read straight from Humboldt's narrative.

The hard-to-find 1969 volume by Adolf Meyer-Abich, extracted from a 1967 monograph in German, is a combined biography and critical assessment of Humboldt's work, with the *Essai* prominently featured. The work provides an excellent summary from a postwar, pre-postmodern, West German perspective. A well-illustrated 1978 biography, intended for a general readership, by Werner Feisst, was published in a dual German-English version.

Nicolaas Rupke, a German historian of science, has recently published a "metabiography"—that is, an analysis of biographies—of Humboldt. This fascinating and provocative book details how German-language biographies have evolved in their emphasis as society and politics have changed. The Humboldt banner has been variously carried by the Prussian Kingdom, the German Empire, the Weimar Republic, the Third Reich, the postwar divided Germanys, and the current reunited, European-Union Germany. And these various periods have all differed in emphasis from the treatments of Humboldt by English, French, North American, and Latin American writers, which in turn vary systematically. Each of these diverse cultures has found something—usually something unique—in Humboldt to admire, idolize, and exploit.

A Humboldt iconography was compiled by Halina Nelken in 1980. This volume includes every known contemporary sketch, painting, bust, relief, and photograph of Humboldt, with descriptions of the circumstances surrounding the images.

The botanical endeavors of the Humboldt/Bonpland voyage are summarized in a 1968 compendium, *Humboldt, Bonpland, Kunth, and Tropical American Botany*, edited by William T. Stearn. This volume reprints a number of useful earlier works, including detailed itineraries for the South American travels reconstructed from notes and manuscripts and from the dates and localities noted in the actual plant collections.

Articles and book chapters on Humboldt are abundant and cover many aspects of his life and work. Contributions by Michael Dettelbach, Anne Marie Claire Godlewska, and Malcolm Nicolson have focused on his botanical, biogeographic, and ecological work.

Many aspects of Humboldt's personality are explored in Daniel Kehlmann's novel, *Measuring the World*, recently translated from German to English. The novel focuses on the contrasting personalities of Humboldt and his contemporary, the mathematician Carl Friedrich Gauss. Each of these luminaries is played off against more-normal men who lived in their shadows—Bonpland for Humboldt, Eugen Gauss for his father. Although liberties are taken with certain facts and details, the novel is entertaining and insightful.

HUMBOLDT'S INFLUENCE

Aaron Sachs's recent tour-de-force, *The Humboldt Current*, provides a broad picture of Humboldt's cultural impact in the nineteenth-century United States. Humboldt's impact on Thoreau, Emerson, and Whitman are respectively discussed in *Seeing New Worlds* and *Emerson's Life in Science*, both by Laura Dassow Walls, and *Walt Whitman's America* by David Reynolds. The first of these (on Humboldt and Thoreau) is especially insightful. Humboldt's influence on Frederic Church is well documented in recent volumes by Kelly and Howat and essays by Edmunds Bunkśe and Stephen Jay Gould. The German scholar Ingo Schwartz has written extensively on Humboldt and the United States.

Chapter 3, "Humboldtian Science," of Susan Faye Cannon's *Science in Culture* is a rich source of insight into Humboldt's thinking and influence on nineteenth-century science. Nicolson, Dettelbach, Godlewska, and others have explored much of the territory initially outlined by Cannon.

Humboldt's influence is part of the lore of Charles Darwin and is discussed in virtually every Darwin biography. The most recent definitive and detailed account is a 1999 essay by Patrick Armstrong, who notes in particular how Darwin's *Journal of Researches* parallels Humboldt's *Personal Narrative* in structure and tone.

HUMBOLDT'S INTELLECTUAL ANTECEDENTS

Kant's physical geography is discussed in a 1958 essay by Hartshorne and a 1970 book by J. A. May. Kant's influence on Humboldt's thinking is discussed in Meyer-Abich (1969) and Malcolm Nicolson's essays.

George Forster is the subject of a 1972 English-language biography and critique by Thomas Saine. His influence on Humboldt is described in the various Humboldt biographies and articles. Forster's *A Voyage Round the World* has been reprinted by University of Hawaii Press, together with a brief biography in the editors' introduction. Forster's account of his 1790 journey with Humboldt across northwestern Europe and England has never been translated into English. Johann Forster's narrative of the Cook voyage has also been reprinted by University of Hawaii Press, with introductory and biographical essays.

No English-language biographies seem to be available for Karl Willdenow. His contributions are summarized in various botanical articles, and his relationship with Humboldt is mentioned in most Humboldt biographies.

Goethe's botanical and other scientific contributions are discussed by Rudolf Magnus in *Goethe as a Scientist* and by Charles Engard in his 1952 introduction to *Goethe's Botanical Writings*. The latter is an invaluable source, compiling all of Goethe's diverse writings on botany and related topics. Stephen Jay Gould provides a brief discussion of Goethe's archetypes in evolutionary context in *The Structure of Evolutionary Theory* and in his essay, "More Light on Leaves."

HUMBOLDT'S CONTEMPORARIES AND PREDECESSORS

Biographical information on Humboldt and others has been compiled from a variety of sources. The Internet is a convenient resource, but information quality can be irregular and errors are propagated by duplication. Its primary value was as an entry point, verified wherever possible using published sources. The *Dictionary of Scientific Biography* served as a final arbiter and information source. Much information on his contemporaries came from the various Humboldt biographies (see above), with the most useful ones being Bruhns, de Terra, Kellner, Meyer-Abich, and Botting. Rupke's metabiography was also useful in identifying aspects of Humboldt missed by these other works. Charles Coulston Gillispie's monumental two-volume work on late eighteenth and early nineteenth-century science in France was an invaluable source of information on Humboldt's French associates and influences.

No English-language biographies of Aimé Bonpland have been published. Information about him can be found in the various Humboldt biographies. George Sarton's 1943 biographical essay is particularly good, with revealing insights into Bonpland's personality and contributions to science.

Francisco Hernández and his sixteenth-century botanical and medical exploration of Mexico are discussed in David Freedberg's marvelous *The Eye of the Lynx*, which also details the tortured publication history of Hernández' manuscripts, notes, and drawings.

English translations of La Condamine's narrative and of Juan and Ulloa's description of the expedition have been published periodically since the late eighteenth century; none are currently in print. A recent book, *The Mapmaker's Wife* by Robert Whitaker, focuses on a series of incidents peripheral to the main expedition. However, nearly half the book is devoted to a well-written and informed account of the origins, execution, and fate of the expedition. The expedition is described in V. W. Von Hagen's *South America Called Them*, which also includes a narrative of Humboldt's travels.

Ken Alder's *The Measure of All Things* provides a lively and insightful account of the efforts of Delambre and Méchain to obtain a precise and objective unit of measurement through geodesy. Freshfield's 1920 *The Life of Horace Benedict de Saussure* emphasizes Saussure's mountaineering accomplishments but also discusses his many scientific contributions. No more recent biographies of Saussure have been published. His important and pioneering scientific work deserves a modern review and assessment.

The Ruiz-Pavón expedition to Peru and Chile is detailed in A. R. Steele's *Flowers for the King*, which also includes information on other early botanical work in South America and the development of the botanical sciences in Spain in the eighteenth century. The botanical explorations of Mociño and Sessé are described in Harold Rickett's *The Royal Botanical Expedition to New Spain*, which also includes valuable information about Cervantes. Iris Engstrand's *Spanish Scientists in the New World* also discusses Mociño and Sessé extensively, as well as Malaspina's expeditions.

Thomas Jefferson's scientific activities and interests are discussed in two outstanding books, Silvio Bedini's *Thomas Jefferson: Statesman of Science* and Bernard Cohen's *Science and the Founding Fathers*. Wasserman (1954) discusses Humboldt's correspondence with Jefferson. Paul Russell Cutright's 1969 *Lewis & Clark: Pioneering Naturalists* focuses on the natural history of the Lewis and Clark Expedition and is one of the finest accounts of the journey. It has been reprinted by University of Nebraska Press. Accounts of the lesser-known southwestern expeditions that Jefferson commissioned are described in volumes edited by Dan Flores (1984) and by Trey Berry and others (2006).

The most comprehensive references on French botany and botanists of the eighteenth and early nineteenth century are *Botanophilia in Eighteenth-Century France* and *French Botany in the Enlightenment*, both by Roger Williams. E. C. Spary's *Utopia's Garden* is also a valuable source.

The life and contributions of Francisco José Caldas are discussed by John Wilton Appel (1994). Jorge Cañizares-Esguerra has recently published an informed and thoughtful discussion of Humboldt's relationship with Caldas and Mutis. Karl Zimmerer discusses Humboldt's relationships with Andean naturalists in a 2006 essay.

Helen Maria Williams's life and works are discussed by Deborah Kennedy in a fine volume, *Helen Maria Williams and the Age of Revolution*. Williams's work with Humboldt is outlined in Nigel Leask's 2001 essay.

Bibliography

This bibliography includes all works specifically referenced in Stephen T. Jackson's essays and appendices, as well as other materials that proved particularly useful in preparation of the volume. It is heavily English-biased and omits most of the vast Humboldt literature published in French, German, and Spanish.

WORKS BY ALEXANDER VON HUMBOLDT

Humboldt and Bonpland, Essay on the Geography of Plants

1807. *Essai sur sur la géographie des plantes*. Paris: Fr. Schoell.

1807. *Ideen zu einer Geographie der Pflanzen nebst einem Naturgemälde der Tropenländer*. Tübingen: F. G. Cotta.

1959. *Essai sur la géographie des plantes (1807)*. Facsimile reprint, Sherborn Fund Facsimile No. 1. London: Society for the Bibliography of Natural History.

1963. *Ideen zu einer Geographie der Pflanzen*. Facsimile reprint, with an essay by Adolf Meyer-Abich. Darmstadt: Wissenschaftliche Buchgesellschaft.

1965. *Essai sur la géographie des plantes*. Facsimile reprint. Mexico City: Institut Panaméricain de Géographie et d'Histoire.

1977. *Essai sur la géographie des plantes*. Facsimile reprint. New York: Arno Press.

1985. *Ideas para una Geografía de las Plantas mas un Cuadro de la Naturaleza de los Paises Tropicales*. Bogotá: Jardin Botanico "José Celestino Mutis."

1990. *Essai sur la géographie des plantes*. Facsimile reprint, with essays by Charles Minguet and Jean Theodorides. Nanterre: Editions Européennes Erasme.

HUMBOLDT, OTHER WORKS

1797. *Versuche über die gereizte Muskel- und Nervenfaser nebst Vermuthungen über den chemische Prozess des Lebens in den Thier- und Pflantzenwelt*. Berlin: Rottmann.

1808. *Ansichten der Natur*. Tübingen: F. G. Cotta.

1811. *Political Essay on the Kingdom of New Spain*. 4 vols. Translated from the original French by John Black. London: Longman, Hurst, Rees, Orme and Brown. Facsimile reprint, New York: AMS Press, 1966.

1814. *Researches Concerning the Institutions and Monuments of the Ancient Inhabitants of America, with Descriptions and Views of some of the most Striking Scenes in the Cordilleras*. 2 vols. Translated by Helen Maria Williams. London: Longman, Hurst, Rees, Orme and Brown, J. Murray and H. Colburn. Facsimile reprint, Amsterdam: Theatrum Orbis Terrarum Ltd. and Da Capo Press, 1972.

And Aimé Bonpland. 1814–29. *Personal Narrative of Travels to the Equinoctial Regions of the New Continent, During the Years 1799–1804*. 7 vols., 1 (1814), 2 (1814), 3 (1818), 4 (1819), 5 (1821), 6 (1826), 7 (1829). Translated by Helen Maria Williams. London: Longman, Hurst, Rees, Orme, and Brown.

1816. Sur les lois que l'on observe dans la distribution des formes végétales. *Annales de chimie et de physique* 1:225–39.

1817. *De Distributione Geographica Plantarum*. Paris: Lutetiæ Parisiorum, in Libraria Græco-Latino-Germanica.

1817. Des lignes isothermes et de la distribution de la chaleur sur le globe. *Mémoires de physique et de chimie de la Société d'Arcueil* 3:462–602.

1820. On Isothermal Lines, and the Distribution of Heat over the Globe. *Edinburgh Philosophical Review* 3:1–20, 256–74; 4:23–37, 262–81; 5:28–39.

1820. On the Cinchona Forests of South America. In *An Illustration of the Genus Cinchona: Comprising Descriptions of all the Officinal Peruvian Barks, Including Several New Species*, edited by Aylmer Bourke Lambert, 19–59. London: John Searle.

1823. *A Geognostical Essay on the Superposition of Rocks, in Both Hemispheres*. London: Longman, Hurst, Rees, Orme, Brown, and Green.

1850. *Aspects of Nature, in Different Lands and Different Climates: With Scientific Elucidations*. Translated by Mrs. [Elizabeth Leeves] Sabine. Philadelphia: Lea and Blanchard.

1850. *Cosmos: A Sketch of the Physical Description of the Universe*. 3 vols. Translated from the 3d German edition (1849) by E. C. Otté. London: Henry G. Bohn; New York: Harper and Brothers. Facsimile reprint of vols. 1 and 2, with introductions by Nicolaas A. Rupke (vol. 1) and Michael Dettelbach (vol. 2). Baltimore: The Johns Hopkins University Press, 1997.

1850. *Views of Nature: Or Contemplations of the Sublime Phenomena of Creation; With Scientific Illustrations*. Translated from the 3d German edition (1849) by E. C. Otté and H. G. Bohn. London: Henry G. Bohn.

1851. *Personal Narrative of Travels to the Equinoctial Regions of America During the Years 1799–1804.* 3 vols. Translated by Thomasina Ross. London: George Routledge and Sons.

1860. *Letters of Alexander von Humboldt to Varnhagen von Ense, From 1827 to 1858.* Translated from the 2d German edition by Friedrich Kapp. London: Sampson Low, Son and Co.

1866. *Cosmos: A Sketch of a Physical Description of the Universe.* Vol. 4. Translated by E. C. Otté and B. H. Paul. London: Henry G. Bohn; New York: Harper and Brothers.

1868. *Tableaux de la nature.* Translated by M. Ch. Galuski. Paris: Librairie Théodore Morgand.

1988. *Political Essay on the Kingdom of New Spain.* The John Black translation, abridged. Edited with an introduction by Mary Maples Dunn. Norman: University of Oklahoma Press.

1990. *Tableaux de la nature.* 2 vols., with essays by Charles Minguet, Philippe Babo, and Jean-Paul Duviols. Nanterre : Editions Européennes Erasme.

1995. *Personal Narrative of a Journey to the Equinoctial Regions of the New Conotinent.* Abridged and translated with an introduction by Jason Wilson, and a historical introduction by Malcolm Nicolson. London: Penguin Books.

PRIMARY AND SECONDARY SOURCES

Ackerknecht, Erwin H. 1955. George Forster, Alexander von Humboldt, and Ethnology. *Isis* 46:83–95.

Agassiz, Louis. 1850. *Lake Superior: Its Physical Character, Vegetation, and Animals, Compared With Those of Other and Similar Regions. With a Narrative of the Tour by J. Elliott Cabot. And Contributions by Other Scientific Gentlemen.* Boston: Gould, Kendall and Lincoln.

————. 1859. Eulogy by Professor Agassiz upon Baron von Humboldt, delivered before the American Academy of Arts and Sciences at their Annual Meeting, on Tuesday, 24 May, 1859. *Littell's Living Age* 61 (no. 785): 643–49.

Agassiz, Louis, and Mrs. Louis. 1868. *A Journey in Brazil.* Boston: Ticknor and Fields.

Alder, Ken. 2002. *The Measure of All Things: The Seven-Year Odyssey and Hidden Error that Transformed the World.* New York: Free Press.

Anonymous. 1882. *Story of the Life and Travels of Alexander von Humboldt.* London: T. Nelson and Sons.

Appel, John Wilton. 1994. *Francisco José de Caldas: A Scientist at Work in Nueva Granada.* Philadelphia: American Philosophical Society.

Armstrong, Patrick. 1999. Charles Darwin's Image of the World: The Influence of Alexander von Humboldt on the Victorian Naturalist. *Beiträge zur Regionalen Geographie* 49:46–53.

Barlow, Nora, ed. 1967. *Darwin and Henslow: The Growth of an Idea. Letters 1831–1860.* Berkeley and Los Angeles: University of California Press.

Bedini, Silvio. 1990. *Thomas Jefferson: Statesman of Science.* New York: Macmillan.

Beniger, James R., and Dorothy L. Robyin. 1978. Quantitative Graphics in Statistics: A Brief History. *American Statistician* 32:1–11.

Berg, Edward E., J. David Henry, Christopher L. Fastie, Andrew D. De Volder, and Steven M. Matsuoka. 2006. Spruce Beetle Outbreaks on the Kenai Peninsula, Alaska, and Kluane National Park and Reserve, Yukon Territory: Relationship to Summer Temperatures and Regional Differences in Disturbance Regimes. *Forest Ecology and Management* 227:219–32.

Berry, Trey, Pam Beasley, and Jeanne Clements, eds. 2006. *The Forgotten Expedition, 1804–1805: The Louisiana Purchase Journals of Dunbar and Hunter.* Baton Rouge: Louisiana State University Press.

Billings, W. Dwight. 1951. Vegetational Zonation in the Great Basin of Western North America. *Comptes rendus du colloque sur les bases écologiques de la régénération de la végétation des zones arides,* 101–22. Paris: IUBS.

Bodenheimer, F. S. 1955. Zimmermann's *Specimen Zoologiae Geographiae Quadripedum*—A Remarkable Zoogeographical Publication of the End of the 18th Century. *Archives Internationales d'Histoire des Sciences* 8:351–57.

Botting, Douglas. 1973. *Humboldt and the Cosmos.* New York: Harper and Row.

Bowen, M. J. 1970. Mind and Nature: The Physical Geography of Alexander von Humboldt. *Scottish Geographical Magazine* 86:222–33.

Bowler, Peter J. 1990. *Charles Darwin: The Man and his Influence.* Cambridge: Cambridge University Press.

———. 1993. *The Norton History of the Environmental Sciences.* New York: W. W. Norton and Company.

———. 2002. Climb Chimborazo and See the World. *Science* 298:63–64.

Bradley, Raymond S., Mathias Vuille, Henry F. Diaz, and Walter Vergara. 2006. Threats to Water Supplies in the Tropical Andes. *Science* 312:1755–56.

Brann, E. R. 1954. *The Political Ideas of Alexander von Humboldt: A Brief Preliminary Study.* Madison, Wis.: Little Printing Co.

Bray, J. Roger, and John T. Curtis. 1957. An Ordination of the Upland Forest Communities of Southern Wisconsin. *Ecological Monographs* 27:325–49.

Breshears, David D., Neil S. Cobb, Paul M. Rich, Kevin P. Price, Craig D. Allen, Randy G. Balice, William H. Romme, Jude H. Kastens, M. Lisa Floyd, Jayne Belnap, Jesse J. Anderson, Orrin B. Myers, and Clifton W. Meyer. 2005. Regional Vegetation Die-off in Response to Global-change-type Drought. *Proceedings of the National Academy of Sciences* 102:15144–48.

Brock, W. H. 1994. Humboldt and the British: A Note on the Character of British Science. *Annals of Science* 50:365–72.

Browne, Janet. 1983. *The Secular Ark: Studies in the History of Biogeography.* New Haven: Yale University Press.

Browne, Janet. 1995. *Charles Darwin: Voyaging. Part One of a Biography.* New York: Alfred A. Knopf.

Bruhns, Karl, ed. 1873. *Life of Alexander von Humboldt.* Compiled in Commemoration of the Centenary of his Birth by J. Löwenberg, Robert Ave-Lallemant, and Alfred Dove. 2 vols. London: Longmans, Green, and Co.

Bunkśe, Edmunds V. 1981. Humboldt and An Aesthetic Tradition in Geography. *Geographical Review* 71:127–46.

Bush, Mark B., and John R. Flenley, eds. 2007. *Tropical Rainforest Responses to Climatic Change.* Berlin: Springer.

Bush, Mark B., and Miles R. Silman. 2007. Amazonian Exploitation Revisited: Ecological Asymmetry and the Policy Pendulum. *Frontiers in Ecology and the Environment* 5:457–65.

Bush, Mark B., Miles R. Silman, Mauro B. de Toledo, Claudia Listopad, William D. Gosling, Christopher Williams, Paulo E. de Oliveira, and Carolyn Krisel. 2007. Holocene Fire and Occupation in Amazonia: Records from Two Lake Districts. *Philosophical Transactions of the Royal Society B* 362:209–18.

Bynum, W. F., E. J. Browne, and Roy Porter, eds. 1981. *Dictionary of the History of Science.* Princeton: Princeton University Press.

Caldas, Francisco José de. 1966. *Obras Completas de Francisco Jose de Caldas.* Bogotá: Universidad Nacional de Colombia.

Candolle, Alphonse de. 1855. *Géographie botanique raisonnée.* 2 vols. Paris : Librairie de Victor Masson.

———. 1885. *Origin of Cultivated Plants.* International Scientific Series, vol. 48. New York: D. Appleton and Co.

Cannon, Susan Faye. 1978. *Science in Culture: The Early Victorian Period.* New York: Dawson and Science History Publications.

Cañizares-Esguerra, Jorge. 2005. How Derivative was Humboldt? Microcosmic Nature Narratives in Early Modern Spanish America and the (Other) Origins of Humboldt's Ecological Sensibilities. In *Colonial Botany: Science, Commerce, and Politics in the Early Modern World,* edited by Londa Schiebinger and Claudia Swan, 148–65. Philadelphia: University of Pennsylvania Press.

Clapperton, Chalmers M., and Colin McEwan. 1985. Late Quaternary Moraines in the Chimborazo Area, Ecuador. *Arctic and Alpine Research* 17:135–42.

Cohen, Bernard. 1995. *Science and the Founding Fathers: Science in the Political Thought of Thomas Jefferson, Benjamin Franklin, John Adams and James Madison.* New York: W. W. Norton.

Connell, Joseph H. 1961. The Influence of Interspecific Competition and Other Factors on the Distribution of the Barnacle *Chthamalus stellatus*. *Ecology* 42:710–23.

Coudrain, Anne, Bernard Francou, and Zbigniew W. Kundzewicz. 2005. Glacier Shrinkage in the Andes and Consequences for Water Resources—Editorial. *Hydrological Sciences Journal* 50:925–32.

Cox, John D. 2005. *Climate Crash: Abrupt Climate Change and What It Means for Our Future*. Washington, D.C.: Joseph Henry Press.

Cutright, Paul Russell. 1969. *Lewis & Clark: Pioneering Naturalists*. Champaign/ Urbana: University of Illinois Press.

Cuvier, M. le Baron G. 1830. *Discours sur les révolutions de la surface du globe*. 6th ed. Paris: Edmond D'Ocagne.

Darwin, Charles. 1897. *The Life and Letters of Charles Darwin, Including an Autobiographical Chapter*. 2 vols. New York: D. Appleton and Co.

———. [1868] 1897. *The Variation of Animals and Plants Under Domestication*. 2 vols. New York: D. Appleton and Co.

———. 1903. *More Letters of Charles Darwin: A Record of his Work in a Series of Hitherto Unpublished Letters*. Edited by Francis Darwin and A. C. Seward. 2 vols. London: John Murray.

De Terra, Helmut. 1955. *Humboldt: The Life and Times of Alexander von Humboldt 1769–1859*. New York: Alfred A. Knopf.

Desmond, Adrian, and James Moore. 1991. *Darwin*. New York: Time Warner.

Dettelbach, Michael. 1996. Humboldtian Science. In *Cultures of Natural History*, edited by N. Jardine, J. A. Secord and E. C. Spary, 287–304. Cambridge: Cambridge University Press.

———. 1996. Global Physics and Aesthetic Empire: Humboldt's Physical Portrait of the Tropics. In *Visions of Empire: Voyages, Botany, and Representations of Nature*, edited by David Philip Miller and Peter Hanns Reill, 258–92. Cambridge: Cambridge University Press.

———. 1999. The Face of Nature: Precise Measurement, Mapping, and Sensibility in the Work of Alexander von Humboldt. *Studies in History and Philosophy of Biological and Biomedical Sciences* 30:473–504.

———. 2001. Alexander von Humboldt between Enlightenment and Romanticism. *Northeastern Naturalist*, Special Issue 1:9–20.

———. 2005. The Stimulations of Travel: Humboldt's Physiological Construction of the Tropics. In *Tropical Visions in an Age of Empire*, edited by Felix Driver and Luciana Martins, 43–85. Chicago: University of Chicago Press.

Drouin, Jean-Marc. 1997. Botanical Geography. In *The European Origins of Scientific Ecology (1800–1901)*, edited by Pascal Acot, 9–18. Amsterdam: Gordon and Breach.

Egerton, Frank N. 1970. Humboldt, Darwin, and Population. *Journal of the History of Biology* 3:325–60.

Emerson, Ralph Waldo. 1884. An Abstract of Mr. Emerson's Remarks Made at the Celebration of the Centennial Anniversary of the Birth of Alexander von Humboldt, September 14, 1869. In *The Complete Works of Ralph Waldo Emerson*, vol. 11: *Miscellanies (1884)*, part 24. Boston: Houghton Mifflin Co..

Engstrand, Iris H. W. 1981. *Spanish Scientists in the New World: The Eighteenth-Century Expeditions*. Seattle: University of Washington Press.

Feisst, Werner. 1978. *Alexander von Humboldt, 1769–1859: Das Bild seiner Zeit in 200 zeitgenössischen Stichen*. Wuppertal: Dr. Wolfgang Schwartze Verlag.

Flannery, Tim. 2005. *The Weather Makers*. New York: Atlantic Monthly Press.

Flores, Dan L., ed. 1984. *Jefferson and Southwestern Exploration: The Freeman and Custis Accounts of the Red River Expedition of 1806*. Lincoln: University of Nebraska Press.

Forster, George. 2000. *A Voyage Round the World*. Edited by Nicholas Thomas and Oliver Berghof, assisted by Jennifer Newell. Honolulu: University of Hawaii Press.

Forster, Johann Reinhold. 1996. *Observations Made During a Voyage Round the World*. Edited by Nicholas Thomas, Harriet Guest, and Michael Dettelbach. Honolulu: University of Hawaii Press.

Freedberg, David. 2002. *The Eye of the Lynx: Galileo, his Friends, and the Beginnings of Modern Natural History*. Chicago: University of Chicago Press.

Freshfield, Douglas W. 1920. *The Life of Horace Benedict de Saussure*. London: Edward Arnold.

Friis, Herman R. 1960. Baron Alexander von Humboldt's Visit to Washington, D.C., June 1 through June 13, 1804. *Records of the Columbia Historical Society, Washington, D.C.* 1960–1962:1–35.

Gillispie, Charles C. 1980. *Science and Polity in France at the End of the Old Regime*. Princeton: Princeton University Press.

———. 2004. *Science and Polity in France: The Revolutionary and Napoleonic Years*. Princeton: Princeton University Press.

Glacken, Clarence J. 1967. *Traces on the Rhodian Shore: Nature and Culture in Western Thought from Ancient Times to the End of the Eighteenth Century*. Berkeley and Los Angeles: University of California Press.

Godlewska, Anne Marie Claire. 1999. From Enlightenment Vision to Modern Science? Humboldt's Visual Thinking. In *Geography and Enlightenment*, edited by David N. Livingstone and Charles W.J. Withers, 236–75. Chicago: University of Chicago Press.

Godlewska, Anne Marie Claire. 1999. *Geography Unbound: French Geographic Thought from Cassini to Humboldt*. Chicago: University of Chicago Press.

Goethe, Johann Wolfgang von. 1952. *Goethe's Botanical Writings*. Translated by Bertha Mueller. Honolulu: University of Hawaii Press.

Goetzmann, William H. 1959. *Army Exploration in the American West, 1803–1863*. New Haven: Yale University Press.

———. 1979. Paradigm Lost. In *The Sciences in the American Context: New Perspectives*, edited by Nathan Reingold, 21–34. Washington, D.C.: Smithsonian Institution Press.

———. 1986. *New Lands, New Men: America and the Second Great Age of Discovery*. New York: Viking.

Goetzmann, William H., and Kay Sloan. 1982. *Looking Far North: The Harriman Expedition to Alaska, 1899*. Princeton: Princeton University Press.

Gould, Stephen Jay. 1989. Church, Humboldt, and Darwin: The Tension and Harmony of Art and Science. In *Frederic Edwin Church*, edited by Franklin Kelly, 94–107. Washington, D.C.: National Gallery of Art and Smithsonian Institution Press.

———. 1993. More Light on Leaves. In *Eight Little Piggies: Reflections in Natural History*, 153–65. New York: W. W. Norton.

———. 2001. *The Structure of Evolutionary Theory*. Cambridge, Mass.: Harvard University Press.

———. 2002. Art Meets Science in *The Heart of the Andes*: Church Paints, Humboldt Dies, Darwin Writes, and Nature Blinks in the Fateful Year of 1859. In *I Have Landed: The End of a Beginning in Natural History*, 90–109. New York: Harmony Books.

Gray, Asa. 1859. Brown and Humboldt. *Proceedings of the American Academy of Arts and Sciences 4*. Reprinted in *Scientific Papers of Asa Gray*, selected by Charles Sprague Sargent, 2:283–88. Boston: Houghton Mifflin and Co.

Hagen, Joel B. 1992. *An Entangled Bank: The Origins of Ecosystem Ecology*. New Brunswick, N.J.: Rutgers University Press.

Hartshorne, Richard. 1958. The Concept of Geography as a Science of Space, from Kant and Humboldt to Hettner. *Annals of the Association of American Geographers* 48:97–106.

Helferich, Gerard. 2004. *Humboldt's Cosmos*. New York: Gotham Books.

Hessenbruch, Arne. 2000. *Reader's Guide to the History of Science*. London: Fitzroy Dearborn Publishers.

Hicke, Jeffrey A., Jesse A. Logan, James Powell, and Dennis S. Ojima. 2006. Changing Temperature Influences Suitability for Modeled Mountain Pine Beetle (*Dendroctonus ponderosae*) outbreaks in the western United States. *Journal of Geophysical Research* 111, G02019, doi:10.1029/2005JG000101.

Hooghiemstra, Henry, and Thomas Van der Hammen. 2004. Quaternary Ice-Age Dynamics in the Colombian Andes: Developing an Understanding of our Legacy. *Philosophical Transactions of the Royal Society of London B* 359:173–81.

Hooker, Joseph Dalton. 1854. *Himalayan Journals, or Notes of a Naturalist in Bengal, the Sikkim and Nepal Himalayas, the Khasia Mountains, &c.* 2 vols. London: John Murray.

Hooker, William Jackson. 1837. Geography Considered in Relation to the Distribution of Plants. In *The Encyclopaedia of Geography*, edited by Hugh Murray, 1:236–54. Philadelphia: Carey, Lea and Blanchard.

Howat, John K. 1995. *Frederic Church.* New Haven: Yale University Press.

Hutton, Charles. 1795. *A Mathematical and Philosophical Dictionary.* London: J. Johnson.

Imbrie, John, and Katherine P. Imbrie 1979. *Ice Ages: Solving the Mystery.* Short Hills, N.J.: Enslow Publishers.

Ingersoll, Robert G. 1900. Humboldt: The Universe is Governed by Law. In *The Writings of Robert G. Ingersoll*, vol. 1: *Lectures*, 93–117. New York: Dresden Publishing Co.

IPCC (Intergovernmental Panel on Climate Change). 2007. *Climate Change 2007—The Physical Science Basis: Working Group I Contribution to the Fourth Assessment Report of the IPCC.* Cambridge: Cambridge University Press.

Jackson, Stephen T. 2004. Late Quaternary Biogeography: Linking Biotic Responses to Environmental Variability across Timescales. In *Frontiers of Biogeography*, edited by Mark Lomolino and Lawrence Heaney, 47–64. Sunderland, Mass.: Sinauer Associates.

———. 2006. Vegetation, Environment, and Time: The Origination and Termination of Ecosystems. *Journal of Vegetation Science* 17:549–57.

Jackson, Stephen T., and Jonathan T. Overpeck. 2000. Responses of Plant Populations and Communities to Environmental Changes of the Late Quaternary. *Paleobiology* 26 (Supplement): 194–220.

Kehlmann, Daniel. 2006. *Measuring the World.* Translated by Carol Brown Janeway. New York: Pantheon.

Kellner, L. 1963. *Alexander von Humboldt.* Oxford: Oxford University Press.

Kelly, Franklin. 1988. *Frederic Edwin Church and the National Landscape.* Washington, D.C.: Smithsonian Institution Press.

———, ed. 1989. *Frederic Edwin Church.* Washington, D.C.: National Gallery of Art and Smithsonian Institution Press.

Kennedy, Deborah. 2002. *Helen Maria Williams and the Age of Revolution.* Lewisburg, Penn.: Bucknell University Press.

Kerr, Richard A. 2007. Pushing the Scary Side of Global Warming. *Science* 316:1412–13.

Kettenmann, Helmut. 1997. Alexander von Humboldt and the Concept of Animal Electricity. *Trends in Neuroscience* 20:239–42.

Kingsland, Sharon K. 1985. *Modeling Nature: Episodes in the History of Population Ecology.* Chicago: University of Chicago Press.

———. 2005. *The Evolution of American Ecology 1980–2000*. Baltimore: Johns Hopkins University Press.

Klencke, Hermann. 1852. *Alexander von Humboldt: A Biographical Monument*. Translated by Juliette Bauer. London: Ingram, Cooke and Co.

Leask, Nigel. 2001. Salons, Alps, and Cordilleras: Helen Maria Williams, Alexander von Humboldt, and the Discourse of Romantic Travel. In *Women, Writing, and the Public Sphere, 1700–1830*, edited by Elizabeth Eger, Charlotte Grant, Clíona ó Gallchoir and Penny Warburton, 217–35. Cambridge: Cambridge University Press.

Livingstone, David N. 1992. *The Geographical Tradition: Episodes in the History of a Contested Enterprise*. Oxford: Blackwell.

———. 2003. *Putting Science in its Place: Geographies of Scientific Knowledge*. Chicago: University of Chicago Press.

Lomolino, Mark V., Dov F. Sax, and James H. Brown, eds. 2004. *Foundations of Biogeography: Classic Papers With Commentaries*. Chicago: University of Chicago Press.

MacArthur, Robert H. 1958. Population Ecology of Some Warblers of Northeastern Coniferous Forests. *Ecology* 39:599–619.

MacGillivray, W. 1836. *The Travels and Researches of Alexander von Humboldt: Being a Condensed Narrative of his Journeys in the Equinoctial Regions of America, and in Asiatic Russia; Together With Analyses of his More Important Investigations*. 3d ed., rev. Edinburgh: Oliver and Boyd.

———. 1852. *The Travels and Researches of Alexander von Humboldt. Enlarged Edition, with a Narrative of Humboldt's Most Recent Researches, Including his Celebrated Journey to the Ural Mountains, Exploration of the Altaian Range, and the Caspian Sea, &c. &c*. London: T. Nelson and Sons.

Magnus, Rudolf. 1949. *Goethe as a Scientist*. New York: Henry Schuman.

Mann, Charles C. 2005. *1491: New Revelations of the Americas Before Columbus*. New York: Alfred A. Knopf.

May, J. A. 1970. *Kant's Concept of Geography and its Relation to Recent Geographical Thought*. Toronto: University of Toronto Press.

McIntosh, Robert P. 1985. *The Background of Ecology: Concept and Theory*. Cambridge: Cambridge University Press.

Merriam, C. Hart. 1890. *Results of a Biological Survey of the San Francisco Mountain Region and Desert of the Little Colorado, Arizona*. Washington, D.C.: United States Department of Agriculture, Division of Ornithology and Mammalogy, North American Fauna No. 3.

Merton, Robert K. 1993. *On the Shoulders of Giants: A Shandean Postscript. The Post-Italianate Edition*. Chicago: University of Chicago Press.

Meyer-Abich, Adolf. 1969. *Alexander von Humboldt 1769/1969*. Bonn: Internationes.

Morley, Robert J. 2000. *Origin and Evolution of Tropical Rain Forests.* Chichester: John Wiley and Sons.

Nelken, Halina. 1980. *Alexander von Humboldt. His Portraits and Their Artists. A Documentary Iconography.* Berlin: Dietrich Reimer Verlag.

Nicolson, Malcolm. 1987. Alexander von Humboldt, Humboldtian Science, and the Origins of the Study of Vegetation. *History of Science* 25:167–94.

———. 1990. Alexander von Humboldt and the Geography of Vegetation. In *Romanticism and the Sciences,* edited by Andrew Cunningham and Nicholas Jardine, 169–85. Cambridge: Cambridge University Press.

———. 1996. Humboldtian Plant Geography after Humboldt: The Link to Ecology. *British Journal for the History of Science* 29:289–310.

Parmesan, Camille, and Gary Yohe. 2003. A Globally Coherent Fingerprint of Climate Change Impacts across Natural Systems. *Nature* 421:37–42.

Playfair, William. 2005. *The Commercial and Political Atlas and Statistical Breviary.* Edited and with an introduction by Howard Wainer and Ian Spence. Cambridge: Cambridge University Press.

Pyenson, Lewis, and Susan Sheets-Pyenson. 1999. *Servants of Nature: A History of Scientific Institutions, Enterprises, and Sensibilities.* New York: W. W. Norton and Co.

Raby, Peter. 2001. *Alfred Russel Wallace: A Life.* Princeton: Princeton University Press, Princeton.

Rasmussen, D. Irvin. 1941. Biotic Communities of Kaibab Plateau, Arizona. *Ecological Monographs* 11:229–75.

Raup, Hugh M. 1942. Trends in the Development of Geographic Botany. *Annals of the Association of American Geographers* 32:319–53.

Reynolds, David S. 1995. *Walt Whitman's America: A Cultural Biography.* New York: Alfred A. Knopf.

Rickett, Harold W. 1947. *The Royal Botanical Expedition to New Spain.* Chronica Botanica, vol. 11, no. 1:1–86. Waltham, Mass.

Rippy, J. Fred, and E. R. Brann. 1947. Alexander von Humboldt and Simón Bolívar. *American Historical Review* 52:697–703.

Robinson, A. H., and Helen M. Wallis. 1967. Humboldt's Map of Isothermal Lines: A Milestone in Thematic Cartography. *Cartographic Journal* 4:119–23.

Romero, Aldemaro, and Kelly M. Paulson. 2001. Humboldt's Alleged Subterranean Fish from Ecuador. *Journal of Spelean History* 36, no. 2:56–59.

Root, Terry L., Jeff T. Price, Kimberly R. Hall, Stephen H. Schneider, Cynthia Rosenzweig, and J. Alan Pounds. 2003. Fingerprints of Global Warming on Wild Animals and Plants. *Nature* 421:57–60.

Ruddiman, William F. 2007. *Earth's Climate: Past and Future.* 2d ed. New York: W. H. Freeman and Co.

Rudwick, Martin J. S. 1997. *Georges Cuvier, Fossil Bones, and Geological Catastrophes: New Translations and Interpretations of the Primary Texts.* Chicago: University of Chicago Press.

Rupke, Nicolaas A. 1999. A Geography of Enlightenment: The Critical Reception of Alexander von Humboldt's Mexico Work. In *Geography and Enlightenment*, edited by David N. Livingstone and Charles W.J. Withers, 319–39. Chicago: University of Chicago Press.

———. 2005. *Alexander von Humboldt: A Metabiography.* Frankfurt/Main: Peter Lang.

Sabrier, Jean-Claude. 1993. *Longitude at Sea in the Time of Louis Berthoud and Henri Motel.* Geneva: Editions Antiquorum.

Sachs, Aaron. 2006. *The Humboldt Current: Nineteenth-Century Exploration and the Roots of American Environmentalism.* New York: Viking.

Sarton, George. 1943. Aimé Bonpland. *Isis* 34:385–99.

Saine, Thomas P. 1972. *Georg Forster.* New York: Twayne Publishers.

Schwartz, Ingo. 1997. The Second Discoverer of the New World and the First American Literary Ambassador to the Old World: Alexander von Humboldt and Washington Irving. *Acta historica Leopoldina* 27:89–97.

———. 2001. Alexander von Humboldt's Visit to Washington and Philadelphia, his Friendship with Jefferson, and his Fascination with the United States. *Northeastern Naturalist* 8 (Special Issue): 43–56.

Serje, Margarita. 2005. The National Imagination in New Granada. In *Alexander von Humboldt: From the Americas to the Cosmos*, edited by Raymond Erickson, Mauricio A. Font and Brian Schwartz, 83–97. New York: Bildner Center for Western Hemisphere Studies, City University of New York.

Smith, Jacqueline A., Geoffrey O. Seltzer, Donald T. Rodbell, and Andrew G. Klein. 2005. Regional Synthesis of Last Glacial Maximum Snowlines in the Tropical Andes, South America. *Quaternary International* 138/139:145–67.

Sobel, Dava. 1995. *Longitude: The True Story of a Lone Genius Who Solved the Greatest Scientific Problem of His Time.* New York: Walker Publishing Co.

Spary, E. C. 2000. *Utopia's Garden: French Natural History from Old Regime to Revolution.* Chicago: University of Chicago Press.

Stearn, William T. 1960. Humboldt's 'Essai sur la géographie des plantes.' *Journal of the Society for the Bibliography of Natural History* 3:351–57.

———. 1968. *Humboldt, Bonpland, Kunth, and Tropical American Botany: A Miscellany on the 'Nova Genera et Species Plantarum.'* Lehre: Verlag von J. Cramer.

Steele, Arthur Robert. 1964. *Flowers for the King: The Expedition of Ruiz and Pavon and the Flora of Peru.* Durham, N.C.: Duke University Press.

Taylor, Bayard. 1859. *The Life Travels and Books of Alexander von Humboldt.* New York: Rudd and Carleton.

Taylor, Steve W., Alan L. Carroll, Rene I. Alfaro, and Les Safranyik. 2006. Forest, Climate and Mountain Pine Beetle Outbreak Dynamics in Western Canada. In *The Mountain Pine Beetle: A Synthesis of Biology, Management, and Impacts on Lodgepole Pine*, edited by Les Safranyik and Bill Wilson, 67–94. Victoria, B.C.: Natural Resources Canada, Canadian Forest Service, Pacific Forestry Centre.

Théodoridès, Jean. 1966. Humboldt and England. *British Journal for the History of Science* 3:39–55.

Thompson, Lonnie G., Ellen Moseley-Thompson, Henry Brecher, Mary Davis, Blanca Léon, Don Les, Ping-Nan Lin, Tracy Mashiotta, and Keith Mountain. 2006. Abrupt Tropical Climate Change: Past and Present. *Proceedings of the National Academy of Sciences* 103:10536–43.

Villegas, Benjamin, ed. 1994. *The Route of Humboldt: Colombia and Venezuela*. Bogotá, Colombia: Villegas Editores.

Von Hagen, Victor Wolfgang. 1945. *South America Called Them*. New York: Alfred A. Knopf.

Wallace, Alfred Russel. 1905. *My Life: A Record of Events and Opinions*. 2 vols. London: Chapman and Hall.

Walls, Laura Dassow. 1995. *Seeing New Worlds: Henry David Thoreau and Nineteenth-Century Natural Science*. Madison: University of Wisconsin Press.

———. 2001. "Hero of knowledge, be our tribute thine": Alexander von Humboldt in Victorian America. *Northeastern Naturalist* 8 (Special Issue): 121–34.

———. 2003. *Emerson's Life in Science: The Culture of Truth*. Ithaca, N.Y.: Cornell University Press.

Wasserman, Felix M. 1954. Six Unpublished Letters of Alexander von Humboldt to Thomas Jefferson. *Germanic Review* 29:191–200.

Whipple, F. J. W. 1921. Discussion on the Theory of the Hair Hygrometer. *Proceedings of the Physical Society* 34:1–1v.

Westerling, A. L., H. G. Hidalgo, D. R. Cayan, and T. C. Swetnam. 2006. Warming and Earlier Spring Increase Western U.S. Forest Wildfire Activity. *Science* 313:940–43.

Whitaker, Robert. 2004. *The Mapmaker's Wife: A True Tale of Love, Murder, and Survival in the Amazon*. New York: Basic Books.

Whittaker, Robert H. 1952. A Study of Summer Foliage Insect Communities in the Great Smoky Mountains. *Ecological Monographs* 22:1–44.

———. 1956. Vegetation of the Great Smoky Mountains. *Ecological Monographs* 26:1–80.

Willdenow, D. C. 1805. *The Principles of Botany, and of Vegetable Physiology*. Translated from the German. Edinburgh: Edinburgh University Press.

———. 1811. *The Principles of Botany, and of Vegetable Physiology*. Translated from the German. A New Edition, Greatly Enlarged by the Author. Edinburgh: Edinburgh University Press.

Williams, John W., and Stephen T. Jackson. 2007. Novel Climates, No-analog Communities, and Ecological Surprises. *Frontiers in Ecology and the Environment* 5:475–82.

Williams, John W., Stephen T. Jackson, and John E. Kutzbach. 2007. Projected Distributions of Novel and Disappearing Climates by 2100 AD. *Proceedings of the National Academy of Sciences* 104:5738–42.

Williams, Roger L. 2001 *Botanophilia in Eighteenth-Century France: The Spirit of the Enlightenment.* Dordrecht: Kluwer.

———. 2003. *French Botany in the Enlightenment: The Ill-Fated Voyages of La Perouse and his Rescuers.* Dordrecht: Kluwer.

Worster, Donald. 1977. *Nature's Economy: The Roots of Ecology.* San Francisco: Sierra Club Books.

Zimmerer, Karl S. 2006. Humboldt's Nodes and Modes of Interdisciplinary Environmental Science in the Andean World. *Geographical Review* 96:335–60.